Enzymes and Enzyme Therapy

D0708654

Other books by Anthony J. Cichoke, M.A., D.C.:

The Back Pain Bible
Bromelain: The Active Enzyme That Helps Us Make the Most of What We Eat (A Keats Good Health Guide)
The Complete Book of Enzyme Therapy
Enzymes: Nature's Energizer (A Keats Good Health Guide)
Introduction to Chiropractic Health: Achieving the Body Balance That Can Defend Health and Protect Against Chronic and Acute Diseases, 3rd edition

Enzymes and Enzyme Therapy

Second Edition

How to Jump-Start Your Way to Lifelong Good Health

DR. ANTHONY J. CICHOKE

KEATS PUBLISHING

LOS ANGELES

NTC/Contemporary Publishing Group

Library of Congress Cataloging-in-Publication Data

Cichoke, Anthony J.
Enzymes and enzyme therapy : how to jump-start your way to lifelong good health / Anthony J. Cichoke.—2nd ed.
p. cm.
Includes bibliographical references.
ISBN: 0-658-00290-2
1. Enzymes—Therapeutic use. I. Title.

RM666.E55C45 2000
615'.35—dc21 99-088463

Published by Keats Publishing
A division of NTC/Contemporary Publishing Group, Inc.
4255 West Touhy Avenue, Lincolnwood (Chicago), Illinois 60712, U.S.A.

Copyright © 1994, 2000 by Anthony J. Cichoke, M.A., D.C.

All rights reserved. No part of this work may be reproduced, stored in a retrieval system, or transmitted in any form or by any means electronic, mechanical, photocopying, recording, or otherwise without prior permission of NTC/Contemporary Publishing Group, Inc.

Printed in the United States of America

International Standard Book Number: 0-658-00290-2

00 01 02 03 04 VP 18 17 16 15 14 13 12 11 10 9 8 7 6 5 4 3 2 1

This book is dedicated to that pioneer in enzyme therapy, the late Professor Doctor Max Wolf. He was a man of dedication, vision, and compassion. His life's work was a fight against killer diseases. Although Max Wolf died on June 3, 1976, his memory will never die. It lives on in his research, in his enzymes, and in the thousands to whom he has given new hope, new life.

Repeatedly, he took abuse from those skeptics who would suppress him and deny his vision. Every day, new research findings confirm Dr. Wolf's insights in the treatment of cancer, cardiovascular disorders, arthritis, multiple sclerosis, herpes, and other deadly disorders. He was not even alive when HIV and AIDS reared their ugly heads, yet his enzymes are being used today to fight these dreaded diseases.

Thank you, Professor Doctor Max Wolf, for bringing a better life to us all through enzymes and enzyme therapy!

Contents

Preface by Abram Hoffer, M.D., Ph.D. ix
Acknowledgments xiii
Introduction: What This Book Can Do for You xv

Part One FUELING YOUR MACHINE WITH ENZYMES

1. Energize Yourself 3
2. Jump-Start Digestion 13
3. Enemies of Digestion 31
4. Toxic Effects 46
5. Garbage in—Garbage out 55

Part Two ENERGIZING ENZYME TOOLS

6. Enzymes' Coenergizers: Vitamins and Minerals 65
7. How to Get Enzymes in Your Life 85
8. The Five-Step Jump-Start Plus Enzyme Program 102

Part Three THE FOUNTAIN OF YOUTH

9. Your Fountain of Youth 113
10. Beauty Is Only Skin Deep 132
11. Bounce Back Faster from Injuries 136
12. Chronic Condition Buster and Immune System Booster 151
13. More Precious Than Gold 166
14. Oh, My Aching Back! 181
15. Enzyme Cancer Fighters 188
16. For Women Only 198
17. Of Interest to Men 205
18. Circulation Rejuvenator 208

19. Victory over Viruses 225
20. New Hope for AIDS 233
21. The Scars of Multiple Sclerosis 240
22. Enzyme Grab Bag 249
23. Our Future with Enzyme Therapy 256

Appendix A 261
Appendix B 267
Appendix C 271
Resource List 273
Bibliography 283
Index 291

Preface

Over the past two years, I have reviewed several books written by A. J. Cichoke, M.A., D.C., describing the use of systemic enzymes for chronic diseases including AIDS, inflammatory lesions, neurological conditions, and other chronic conditions. Enzymes are used in medicine, but not to any large degree and usually only for conditions where it can be established or has been established that there are enzyme deficiencies. Thus, if the pancreas is not functioning properly it makes good medical sense to give that person enough pancreatic enzymes to make up the deficiency. The key word is deficiency. But when there is no reason to believe there is a deficiency, the enzymes are not used.

In this book, Dr. Cichoke describes the use of enzymes in larger quantities for conditions that are not known to be caused by enzyme deficiencies. He describes a paradigm change from using enzymes for deficiencies *only*, to using them for nondeficient conditions as well. It is similar to the changing utilization of vitamins. They were originally used only for prevention of deficiencies, and only latterly have come into more common use for the treatment of conditions that are not vitamin deficiencies.

Enzymes have been used much more commonly by chiropractors, naturopaths, and by physicians in Europe to a much larger degree than they have on this continent. Unfortunately, the European studies have been published in journals that are not available to North American physicians or are not in English so that they could not be read. According to the massive bibliography assembled by Dr. Cichoke, there is a very large European literature that supports the use of enzymes, which he has described in this book.

Another reason why enzymes were not used systemically was the belief that the enzyme molecules are too large to pass through the intestinal walls, that they would be digested, that is broken down, and therefore could not be of any value in the body except in the digestive tract. Like so many old beliefs based upon theoretical notions and not upon experimental evidence, this is not true. For the past twenty years more evidence has been found that these large enzyme molecules can be absorbed intact into the bloodstream and can be taken to other parts of the body in the same way large polypeptide fragments can be absorbed and create havoc in certain organ systems. However, the enzymes do not create havoc, perhaps because the body has been producing enzymes for as long as animals have been digesting food, and the digestive tract and the circulatory system are as familiar with them as they are with vitamins and minerals.

It is curious that enzymes have been neglected for such a long time. Every or nearly every reaction in the body is accelerated or catalyzed by enzymes. They are the catalysts for life, and they are specific for each reaction that demands its own enzymes and support from other molecules called coenzymes and from the other natural constituents of the body, vitamins and minerals.

In introducing the systemic use of these enzymes, Dr. Cichoke joins the ranks of some of our pioneers in orthomolecular medicine such as Dr. O. Truss, who showed the vast importance of chronic yeast infections (Candida) in causing disease; Drs. M. Mandell and W. H. Philpott, who introduced orthomolecular psychiatry to the enormous importance of allergies in generating psychiatric diseases; and Dr. C. C. Pfeiffer, who showed us how essential were some of the minerals, especially zinc and copper, as well as provided for the first time a biochemical classification for the schizophrenias.

This book may introduce to modern medicine the proper use of the enzymes. Fortunately they are readily available in health-food stores, in some drug stores and are available over-the-counter. This means they can be used widely, and since they can do no harm except for the rare allergic reactions, which any organic substance can cause, they can be tried out for the large number of conditions so well described in this book. It also provides the information about enzymes, their function in the body, how they are used, and the clinical results one can expect if they are used as described. I have been so immersed in the use of optimum quantities of vitamins and minerals for conditions not caused by deficiencies, that I too have neglected to use systemic enzyme therapy. However, encouraged by the previous publications written by Dr. Cichoke, and most recently by this book, I have already begun to add them to my orthomolecular practice. For treating conditions ranging from inflammations to cancers, they surely are going to play a major role in future orthomolecular practice.

Abram Hoffer, M.D., Ph.D.

Acknowledgments

The author wishes to thank the following researchers and physicians for their cooperation, advice, assistance, and information in writing this book: Raul Ahumada, Motoyuki Amano, Takahiko Amano, Vladimir Badmaev, Wayne Battenfield, W. Bartsch, D.J.A. Cole, Tony Collier, Michael Culbert, F. W. Dittmar, James Duke, Charles Fox, Naritada Fujiki, G. Gallacchi, G. Gebert, Wilhelm Glenk, Yoshihide Hagiwara, John Heinerman, A. Hoffer, Rudolph Inderst, Hans Jager, Peter Karnezos, Leslie Kenton, Shigeki Kimura, Franz Klaschka, Gert Klein, Michael W. Kleine, Stephen F. Langer, Benjamin Lau, Joe Lehmann, Robert I. Lin, D. A. Lopez, T. Pearse Lyons, Muhammed Majeed, E. J. Menzel, Mark Messina, John Mills, Earl Mindell, Daniel B. Mowrey, Michael Murray, Tracey Mynott, Sven Neu, Christine Neuhoffer, Richard A. Passwater, Otto Pecher, Barbara Pfannenschmidt, Joseph Pizzorno, Joan Priestley, H. D. Rahn, Karl Ransberger, Vic Rathi, Corey Resnick, Robert Rosen, W. Scheef, Ed Schuler, Art Sears, J. Seifert, Lendon Smith, G. Stauder, C. Steffen, Peter Streichhan, Koichi Suzuki, Mitsuru Takiura, Steven Taussig, G. P. Tilz, Ron Tominaga, Klaus Uffelmann, Wolf Vogler, W. Von Schaik, R. Michael Williams, Heinrich Wrba, Janet Zand, and finally Ms. Katie Cichoke and William Cichoke.

A huge thanks goes to the libraries of America and especially to Portland, Oregon's West Slope Library for their hard work, support, and neverending cooperation.

Special thanks to my secretary, Mrs. Karen Hood, for all of her assistance in the preparation of this book, including doing research, typing, and editing; and to Peter Hoffman, Claudia McCowan, and Jama Carter of Lowell House/Keats Publishing for their special work as editors.

Finally, my eternal thanks to my wife, Margie, for her neverending support and for enduring the mountains of paper and manuscripts in our house.

INTRODUCTION

What This Book Can Do for You

IN THE six years since the first edition of this book was published, interest in, and use of, enzymes has exploded. Historically, enzyme supplements have been used as digestive aids. However, their other health benefits including fighting inflammation, improving circulatory disorders, reducing pain, fighting viruses, and keeping you younger longer are now being more widely recognized. In the first version of this book, I briefly touched on some future uses of enzymes—uses that are now mainstream, such as enzymes in toothpaste and enzymes to fight sinusitis and improve neurological conditions.

This book is written for everyone, whether 6 or 60, 1 or 100. It's for all of us who get up each morning to face a very competitive world. It's for those of us who want to have that extra edge, who want to feel better, look better, and live longer—without those hated diseases—who want to slim down, tone up, and give added "juice" to our body's machine.

For the most part, Americans are overweight, undernourished, rundown, and polluted empty shells. We are unhealthy—and that's expensive! In 1960, 5 percent of the Gross National Product (GNP) went to medical services. By 1990, that figure had exceeded 12 percent. Nowadays, we spend nearly one trillion dollars every year on health care.[1] Injury alone now costs us over $100 billion annually, cancer over $70 billion, and cardiovascular disease over $135 billion.[2]

You want to be healthy, but are you willing to give up your favorite smokes? That steaming coffee swirling down your gullet? That cola at 10, 2, and 4 (low blood-sugar time)? The three martinis after a long day's work, the recreational "joint" just to knock off the edge? Do you want to sacrifice your favorite fast-food fandango of hamburgers, fries, and a "liquid sugar" drink?

You want to continue eating your favorite juicy steaks, greasy fries, butter-laden croissants, and multi-tiered, cream-filled, empty carbohydrate desserts. You want your feast without the gory consequences.

But at the same time, you want to stay healthy without a parade of drugs, pills, painful surgeries, starvation routines, or monotonous exercises. You still want to look good, feel good, and live longer, lose weight (or stay slim), all while having an endless supply of energy.

Is this possible? Can the old battery be recharged? Will Superman *ever* marry Lois Lane?

Some questions can never be answered, but we do know you can reenergize your life. We know that Gotham City will be safe. This book is a guide to jump-starting your day (and every day for the rest of your life). Follow this plan for a lifetime and you will find those hidden secrets Aladdin might have asked from his Genie.

By following this manual you can recharge your battery, your engine, your body, and your life. This guide is what you always wanted but could not find elsewhere. There are no drugs, no special foods, no complicated back-cracking exercises. The book shows you how to recharge your day with no special effort on your part. It's easy to follow, simple, and best of all, unregimented.

You'll learn: 1) how enzymes can help you increase your energy level, 2) how to get rid of your body's garbage, 3) what foods give you the most enzyme energy, 4) how to relieve pains and aches, and 5) how to build up your resistance to disease. Plus, you can continue eating most of your favorite foods and *still* feel good, look good, and live longer.

The key to body power and vitality is *enzymes*. Enzymes will help your body work more productively with less effort. It's like using STP in your car's engine. This book offers hope for the too fat, too thin, and for those of us with runaway wrinkles, chronic illness, and no pep. This book is for the downtrodden, the weary and wretched.

Though most programs offer tedious group therapy gatherings, stomach-churning drugs, hunger pangs, and agonizing exercises, this program offers a healthy plan and a change of diet to literally jump-start your day.

Reference

1. *Health Care Financing Review*, Washington, D.C.: U.S. Health Care Financing Administration, fall 1997. Retrieved from http://www.hcfa.gov/stats/stats.htm.
2. *Healthy People 2000*, Washington, D.C.: U.S. Department of Health and Human Services, Public Health Service, 1991, 5.

PART ONE

Fueling Your Machine with Enzymes

CHAPTER 1

Energize Yourself

WHAT JUMP-starts *your* day? What gets you going and keeps you going? Do you need a cup of coffee to "kick into gear"? Did you know that caffeine (among other things) raises blood pressure?[1] In addition, caffeine is addictive, and if you decide to give it up, you might experience headaches, drowsiness, and fatigue. Definitely not the path to energy.

Instead of coffee, do you, like millions of Americans, start your day with a can of caramel-colored soda pop? One *cup* of cola contains 100 calories, no protein,[2] and anywhere from 30 to 46 milligrams of caffeine.[1] Did you know that this combination of caffeine and sugar will guarantee caffeine jitters, a sugar high, and a corresponding 10 AM (blood sugar) crash?

Your body craves energy. But what is energy to you? Do you get through a stress-filled work day only to collapse in front of the TV, sleeping through most of your favorite programs?

If you're a single parent, you probably *don't* have enough energy. Do you drag yourself out of bed, plug through a day's work, only to come home to face dinner, dishes, and laundry? Do you run out of steam before your day is finished?

If you're an athlete, you know that having enough energy is critical to performing and competing at your best.

A JEKYLL-AND-HYDE EFFECT

If you're tired, first look to your yo-yo blood-sugar level as the culprit of chronic or periodic fatigue. What does sugar have to do with energy? Ever hear of low blood sugar (hypoglycemia)? When your blood sugar (glucose) level drops, you feel tired—your energy mysteriously disappears. Don't be blinded by claims of increased or instant energy with sugar-laden or caffeine-laced foods and beverages. The rollercoaster ride they provide begins with your blood sugar shooting up in a burst of energy, then plummeting downward. This rapid rise and fall is accompanied by fatigue and loss of energy. The ups and downs of this deadly yo-yo drain you of energy and can affect you emotionally. For a period of time, you're on top of the world and then, suddenly, you hate everyone.

Ideally, complex sugars gradually raise your blood-sugar level. Sugar from fruits and other complex carbohydrates is slowly released into the bloodstream, creating a gradual and prolonged elevation in energy. This is unlike the refined carbohydrate effect, which causes a rapid insulin response. Think of what would happen if you used ether in your car's engine instead of gas. The engine would rev up like crazy for a short period of time and then crash (burn out).

According to the late Adelle Davis, "Refined sugar overstimulates the production of insulin and alkaline digestive juices, interferes with the absorption of proteins, calcium, and many other minerals, and retards the growth of valuable intestinal bacteria."[3] So why eat it? When you eat simple carbohydrates (such as candies and table sugar), your pancreas is overtaxed and ultimately pays the price by collapsing.

Today's fast-food diet, plus the constant pressures of school or job and/or home and family, seem to pull the cork and set free our vital body energy. We can practically feel energy ooze from our bodies. This constant and prolonged state of stress drains energy from our enzymes, as well as every one of our systems, especially our immune system.

We assault our bodies with empty carbohydrate foods, inadequate exercise, and irregular eating habits. We eat on the run, grabbing toxic substances such as alcohol, coffee, cigarettes, drugs, salt, and sugar, plus giving ourselves little personal time and probably not enough sleep. We have become our own assassins. We kill our enzymes and the guardian of good health—our immune system—which leaves us defenseless. Why shouldn't we feel like a load of laundry just squeezed through a ringer? This energy drain leaves us feeling listless, run down, and depressed. That human dynamo has lost its spark.

Do you feel like bouncing out of bed every morning? Supposedly, if you eat "right," get plenty of rest, and exercise regularly, you won't have any troubles. But can Americans in this day and age really eat "right"?

Research says we're a sick society. We can't get a well-balanced diet because of all the preservatives, additives, storage, canning, freezing, drying, and adulteration of our foods. But, even if it's fresh, is our food still alive?

What keeps food alive? Enzymes—energy-rich enzymes.

What Are Enzymes?

Enzymes are substances that occur naturally in all living things, including the human body. If it's an animal or a plant, it has enzymes.

Enzymes are critical for life. At present, researchers have identified more than 3,000 different enzymes in the human body. Every second of our lives these enzymes are constantly changing and renewing, sometimes at an unbelievable rate.

Our body's ability to function, to repair when injured, and to ward off disease is directly related to the strength and numbers of our enzymes. That's why an enzyme deficiency can be so devastating.

All life processes consist of a complex series of chemical reactions. These reactions are referred to as *metabolism*. Enzymes are the catalysts that make metabolism possible. A catalyst is a substance that initiates a chemical reaction, enabling it to proceed under different conditions (such as at a lower temperature) than would otherwise be possible.

Some writers compare enzymes to the clergy and judges who make and break marriages. When you stand in front of the preacher, he acts as a catalyst—he joins you and your spouse, making you a "married couple." He, however, is not changed in the process. Similarly, if you're unlucky enough to ever face divorce, you know that the judge dissolves your partnership, separates the two of you, but, like the preacher, is unchanged by the action. Without the preacher or the judge, you couldn't get married (or divorced).

Enzymes work this same way. Without them, many of the body's chemical reactions would never take place. Without enzymes, there would be no breathing, no digestion, no growth, no blood coagulation, no perception of the senses, and no reproduction. Our bodies contain millions of enzymes, which continually renew, maintain, and protect us. No person, plant, or animal could exist without them.

Unfortunately, most people have never heard about enzymes, and are unaware of their importance in regulating the body and maintaining health. Vital energy (in the body) only becomes available because of the role enzymes play in metabolism. Enzymes are responsible for synthesizing, joining together, and duplicating whole chains of amino acids. You've probably heard of amino acids—they're the components that make up protein. Enzymes are also proteins, consisting of long chains of amino acids that differ in order and number.

Because enzymes are catalysts, their effectiveness can be greatly influenced by their environment. An acid or an alkaline environment will affect their activity, as will temperature, concentration of substrate (the substance upon which they work), coenzymes or cofactors, and inhibitors.

In addition, enzymes are very specific. Each enzyme promotes one type of chemical reaction and one type only. Some enzymes help break down large nutrient molecules (the proteins, carbohydrates, and fats in our foods) into smaller molecules for digestion. Others are responsible for different functions, such as the storage and release of energy or the processes of respiration, reproduction, vision, and so on.

THE BODY'S BATTERY CHARGER

Every activity that occurs within the body involves enzymes: the beating of the heart, the building and repair of tissue, the digestion and absorption of food; it goes on and on. Nothing can take place without energy and energy cannot be used or produced without enzymes. Enzymes are involved in all bodily functions. In fact, the very existence of each living cell depends on complicated chemical reactions that require a constant supply of energy and enzymes. Without energy, cells become disorganized, resulting in illness and death. It is for this reason that the body's energy needs take precedence over all other body requirements.

Cells take energy from the proteins, carbohydrates, and fats that we eat. They do this with the help of enzymes. Before they reach the cell, practically all proteins are converted into amino acids, fats are converted to fatty acids, and carbohydrates are changed to glucose. The cells oxidize these nutrients, releasing large quantities of energy in the process. Inside the cell, oxygen reacts chemically with the nutrients under the influence of certain enzymes, which control the rate of reactions and direct the released energy toward its proper use.

In combination with oxygen, these same foods can release large amounts of energy when burned outside the body in an actual fire. In this instance, the energy is released suddenly and all in the form of heat. For physiological processes, however, the body needs *energy*, not heat. We need energy to cause mechanical muscle movement and other body functions to occur. To produce this energy, chemical reactions must be "coupled" with the systems responsible for these physiological functions. This coupling is achieved through special energy transfer systems and cellular enzymes.

ATP, Our Body's Energizer

Inside the cells, energy is released and used mainly in the form of *adenosine triphosphate* (ATP). ATP is used throughout the cell to energize almost all the intracellular metabolic reactions.[4]

The amount of energy liberated by complete oxidation of food is called *free energy of food oxidation*. This energy can actually be measured and is expressed in terms of *calories per mole of substance*. For example, the amount of free energy liberated by oxidation of one mole of glucose (180 grams) is 686,000 calories.[5] We'll cover more on calories later.

The body needs nutrients in a sufficient supply to produce adequate energy; without them, life ceases.

Because it is widely used by biological systems for *storing* energy, ATP is an extremely important compound. Energy is liberated when ATP is split and energy is required to reform ATP. Thus, those reactions that *use* ATP by converting it to adenosine *diphosphate* (ADP) are *energy-using reactions*, while those that *produce* ATP are *energy-producing reactions*. Usually, a biological reaction is driven by energy obtained from breaking a high-energy bond, such as the one in ATP. And what splits ATP into ADP, thereby liberating a high-energy phosphate bond? An enzyme, *adenosine triphosphatase* (ATPase).

How Cells Get Energy from Nutrients

Metabolism includes all the physical and chemical processes involved in the activities of life. Enzymes are the *means* within the cells by which the building-up and breaking-down processes of metabolism take place.

Nature has devised a brilliant process to supply the constant demand for energy, called *biologic oxidation*. This process allows us to obtain energy from food without burning up body tissue at the same time. Because of the catalytic activity of enzymes, food can be burned at low temperatures, which are compatible with the life of the cell.

The process of biologic oxidation progresses in a step-by-step fashion and has been compared with the orderly progression of a boat through a series of locks in a river, as opposed to plunging over the falls all in one effort. The citric acid cycle (Krebs cycle) oxidizes the end products of digestion. A tremendous amount of energy is liberated, in addition to carbon dioxide and water. A portion of this energy keeps the body warm; the rest is stored temporarily in the high-energy phosphate bonds of ATP.

Often likened to a cocked and loaded gun, ATP is always ready to "fire" and release its cellular energy when needed by the cells. However, this is only a small part of the picture. Every living cell contains ATP. ATP provides the energy needed for repair and building of protoplasm, as well as mechanical and chemical work.

All carbohydrates, proteins, and fats contain stored energy that, by biologic oxidation in the Krebs cycle, can be transferred to ATP. This potential energy value of food is called the caloric, or heat, value. Often abbreviated "Cal.," a large calorie is the amount of heat required to raise the temperature of 1 kilogram (2.2 pounds) of water 1 degree centigrade. If you've ever dieted, you know what a calorie is.

Ninety percent or more of all the carbohydrates you eat are used to form ATP in the cells. Carbohydrates serve as quick sources of energy because they are very easily burned. When oxidized, each gram of carbohydrate releases 4 calories. If you read labels in an attempt to improve your diet, you know that each gram of fat is the equivalent of 9 calories. Thus, fat has a higher caloric value than carbohydrates or protein. Only a comparatively small amount of protein is used to supply energy, yet, on oxidation, each gram of protein releases 4 calories. The oxidation of body protein is of major importance when food intake is severely restricted, as in starvation. Since energy needs take precedence over all others, even cell protoplasm is oxidized to meet the body's needs. Death is the obvious and ultimate result when there is no resumption of food intake.

The energy output needed to keep a person alive (that is, to maintain circulation, respiration, and body temperature) is called *basal metabolism*. Basal metabolism is related to your surface area (about 1,000 calories per square meter of body surface per day).

The basal metabolic rate (BMR) is influenced by several factors. For example, BMR varies with age, is lower in women than in men (the latter being larger and having more muscle tissue), and also varies with nutritional state. Children, because they grow rapidly, usually have a higher rate. If the thyroid gland is overactive (as in Graves disease) the BMR will be elevated beyond the normal limits. If the thyroid gland is underactive, the basal rate will be depressed below the normal limits. Any abnormal activity of the thyroid gland will cause changes in the BMR, since the thyroid hormone plays an important role in determining the rate at which chemical reactions proceed in the cells. The BMR measures only the overall oxygen consumption; it may be increased in congestive heart failure (CHF) and in various other conditions not related to thyroid, such as unrecognized infections.

Since your total caloric requirement depends on the amount of energy you expend when eating, working, or playing (in addition to the basic needs previously described), it is subject to great individual variation. For example, if you sit at a desk all day, you may need only 2,400 calories per day. On the other hand, a man at very hard physical labor may require 6,000 to 8,000 calories to meet his energy requirements. In other words, energy output should be balanced by caloric input. Problems develop when we eat like a heavy laborer, but follow a sedentary lifestyle.

An excessive intake of food, over and above what we burn, is usually the cause of obesity or overweight (which will be covered more extensively in a later chapter). Excess food is stored as fat. If this is the case, you should reduce your caloric intake, balancing your diet to make sure you're getting enough vitamins, minerals, and life-giving enzymes. But you also must restoke your furnace. Make sure you have the enzymes your body needs to properly digest and absorb what you eat, and to properly "gas up" your systems.

Short Circuits

Enzymes are important for your nervous system, too. Nervous system function is regulated by various neurotransmitters, such as serotonin, the catecholamines (dopamine and norepinephrine), and acetylcholine. These neurotransmitters are manufactured by the action of enzymes in the brain on the precursor amino acids, tryptophan, tyrosine, and choline, respectively. Because the brain cannot make adequate quantities of the various precursors, it must obtain these precursors from the bloodstream. But what if your digestive system hasn't properly broken down the protein you eat into its component amino acids, thus leading to a deficiency state in your bloodstream?

WHERE DO YOU GET ENZYMES?

Now that we have established our need for enzymes and their importance in energy production and the nervous system, where do we get them? Traditionally, foods have been the primary source. But did you know that the heat of cooking can kill enzymes? Uncooked foods (such as fruits and vegetables) are usually high in enzyme activity and, fortunately, taste good, too.

In theory, this works (take food—get enzymes). However, in practice, with the magnitude of food additives, preservatives, radiation, long-term storage, canning, freezing, drying, and so on, the actual enzymatic activity level of foods can be grossly reduced. Can daily supplementation (in addition to foods) ensure an adequate supply of enzymes? Read on.

The Solution

What's the solution to an energy drain? Say "yes" to complex carbohydrates and live enzymes, and "no" to deadly caffeine and refined sugar. The unwanted insulin response that is associated with refined carbohydrates is curbed when we eat complex carbohydrates, allowing blood sugar levels to rise at a normal rate. Eating complex carbohydrates, fresh fruits, and vegetables (which are loaded with active enzymes) allows natural sugars to enter the bloodstream slowly, resulting in a more gradual, but longer-lasting buildup of energy.

We can't produce energy without catalysts, and enzymes are those catalysts.

You can't jump-start your day and feel young, with energy and vitality, if your body has lost its enzyme punch. If your get-up-and-go has got up and gone, your wake-up call may go unheard!

Life is similar to walking on a tightrope. Like everything else, there is a beginning and an end on this tightrope called life. As we move along on our journey, we must balance our bodies (this is known as homeostasis) or we can fall off the tightrope before our time, before reaching our scheduled end. This balancing act involves the total body (mind and spirit), yin and yang, temperature, pH, vitamins, minerals, anabolic-catabolic ratio, the oxidation of body cells, and enzymes.

All must be in harmony, and enzymes help us maintain that balance.

Now that you know enzymes help give us energy, how do we get the nutrients from the fields to the fork and into our body machinery? That's the function of digestion and absorption and will be covered in the next chapter.

References

1. Whitney and Rolfes, *Understanding Nutrition*, 6th ed. St. Paul, Minn.: West Publishing, Inc., 1993, 617–618.
2. Ibid., 103.
3. Adelle Davis, *Let's Get Well.* New York: Harcourt, Brace and World, Inc., 1965, 390.
4. Arthur C. Guyton, *Textbook of Medical Physiology*, 8th ed. Philadelphia: W. B. Saunders Co., 1991, 19–20.
5. Ibid., 744.

Chapter 2

Jump-Start Digestion

HERE COMES a forkful of food. Open wide and into the chasm it goes. Your eyes and nose have already checked out the food and told your brain, "Be on the lookout, mouth, something good is coming your way." In response, your mouth begins to water. Saliva and other enzyme juices are secreted and are ready to mix with the food. The tongue tosses the food around (much like cement is tossed around in a cement mixer), positioning it so your grinding teeth can break it down and tear it apart. It is especially important that the cellular membranes covering the fruits and vegetables be broken; in fact, good digestion depends on it.

Your Food Conveyor Belt

Think of your digestive tract as an enclosed assembly line. As food passes along the conveyor belt, various enzymes spray and cover the nutrients, breaking down the food products into an absorbable, usable form for the body.

That thirty-foot-long system of plumbing that runs from the mouth south to the anus is called the gastrointestinal (GI) tract. This GI system includes the mouth, esophagus, stomach, small and large intestines, plus salivary glands, liver, gallbladder, and pancreas. They all play a part in digestion, and a defect in any of them can cause problems.

The purpose of the GI tract is to break down the big pieces of food in our diet (carbohydrates, fats, and proteins) into absorbable nutrients (i.e., sugars, free fatty acids, and amino acids). The gastrointestinal system has specific functions which include: 1) receipt of foods; 2) breakdown of foods into their basic parts; 3) movement of foods through the system; 4) secretion of acid, mucus, bile, digestive enzymes, and other substances; 5) absorption of the various components of food; and 6) elimination of byproducts and toxic waste products.

So, we've got the food (fuel), now what do we do with it?

Without adequate digestion, good health is impossible. Even the best dietary intake will be of little value if your digestive system can't adequately break down the food for bodily use. In addition, without proper breakdown, the body may absorb macromolecules (including whole bacteria), which could lead to infection, intestinal toxemia or irritation, and a number of diseases.

The purpose of the GI system is actually very simple—to extract nutrients from foods, digesting them into units that are small enough to be absorbed, and to eliminate waste products.

WHY WE NEED ENZYMES

But what do enzymes do—and why do we need them for digestion? Enzymes make the digestive system work. They are present in every phase of digestion and without them, you can't adequately digest what you eat. No matter what we eat, we consume mostly proteins, fats, and carbohydrates. In order to convert these three basic food groups into materials our bodies can use, we need to have three enzyme groups—proteolytic, lipolytic, and amylolytic—working to digest our food.

1. **Proteolytic enzymes,** or **proteases,**
   are those which decompose protein. Protein is made up of about twenty common amino acids. (Though the body can make some amino acids, there are nine that we cannot make or cannot make in adequate quantities.) Like every enzyme, each protease works on a specific amino acid. For instance,

the protease, *trypsin*, splits the amino acids lysine and arginine, while *chymotrypsin* splits phenylalanine, tyrosine, and tryptophan.[1]

Table 2–1. Amino Acids

ESSENTIAL AMINO ACIDS	NON-ESSENTIAL AMINO ACIDS
Histidine	Alanine
Isoleucine	Arginine
Leucine	Asparagine
Lysine	Aspartic acid
Methionine	Cysteine
Phenylalanine	Glutamic acid
Threonine	Glutamine
Tryptophan	Glycine
Valine	Proline
	Serine
	Tyrosine

2. **Lipolytic enzymes,** or **lipases,**
decompose fat (lipids). The lipid family includes triglycerides (fats and oils), phospholipids (such as lecithin), and sterols (such as cholesterol).[2]

3. **Amylolytic enzymes,** or **amylases,**
decompose carbohydrates. In the normal human diet, there are several major sources of carbohydrates. These are *sucrose* (a disaccharide, commonly called cane sugar); *lactose* (a disaccharide in milk); *fructose* (a monosaccharide); and *starches* (large polysaccharides present in almost all nonanimal foods and particularly in grains). In addition to these carbohydrates, we also ingest alcohol, amylose, glycogen, lactic acid, pyruvic acid, pectins, and dextrins. Our diets may also contain large amounts of cellulose, which is a carbohydrate. However, because the human digestive tract produces no *cellulase* (the enzyme that breaks down cellulose), it works as a fiber, and not as an energy source.

The Digestive Process

To understand the role of enzymes in digestion, we must try to understand the digestive process. Let's take a look at what happens after we eat a meal.

The physical and chemical breakdown of food begins in the mouth as your teeth break up the food and mix it with saliva. Humans produce 1 to 1½ liters of saliva per day.[3] Salivary enzymes in the mouth begin to digest the starches that you eat. To experience firsthand saliva's action on starch, put a cracker or a piece of bread in your mouth, chew it, and hold it in your mouth for a short time; it will begin to taste sweet. This is because the amylase in your saliva is breaking the starch down into maltose (a sugar).

Saliva is mainly secreted by the submaxillary, sublingual, and parotid glands, and passes through ducts into the mouth.

What does saliva do? Saliva functions in a number of ways:

1. Digestive: The salivary enzyme, amylase (ptyalin) splits starchy foods (such as spaghetti, bread, and macaroni) into smaller molecules of the disaccharide, maltose. However, ptyalin has no action upon cellulose.
2. Lubrication: Saliva lubricates food for swallowing. This is one of the most important functions of saliva, that is, to moisten food, thus enabling it to be rolled into a mass and given a lubricant coating, for easier swallowing.
3. Solvent action: Taste is a chemical reaction. Therefore, all solid substances, to stimulate the taste buds, must be dissolved in saliva.
4. Cleansing action: The constant flow of saliva has a very necessary cleansing effect on the mouth and teeth, keeping them relatively free of food particles, foreign substances, and sloughed-off tissue cells.
5. Moistening and lubricating action: The soft parts of the mouth and lips are kept moist and lubricated.

Chew Your Way to Better Health

Good digestion begins in the mouth and with chewing. In order for the enzymes to do their job, we must chew well. Chewing helps the enzymes in the mouth to attack your food, getting it ready for its trip down the gullet. Even before the food is swallowed, thorough chewing increases digestive tract enzyme production. It also helps make the foods more digestible and exposes more food surface to enzymatic juices. Further, chewing causes an involuntary secretion of pancreatic fluids, intestinal secretions, and bile, vitally needed to break up the food, thus releasing the energy and life-giving nutrients to the body.

The longer you chew your food, the longer the enzymes in your mouth can work on that food, breaking up starches even before they are swallowed. This is all necessary for the breakdown and absorption of foodstuffs (into nutrients) for the body's use.

Plant cells differ from animal cells in that plants have a cell wall, commonly composed of cellulose. The animal cell has a cell membrane, but not a rigid wall. Chewing is critical for digestion of all foods, because enzymes only act on the surface of food particles. The more time spent chewing, the more surface area of the food will be exposed to enzymes, leading to better digestion. Most raw fruits and vegetables have undigestible cellulose membranes, which must be broken before nutrients can be released and the food can be digested. For this reason, chewing is especially important for raw vegetables and fruits.

Chewing rejuvenates your digestive system. It is an integral part of digestion and helps ptyalin do its job. Starches that are not chewed are washed down to the stomach with liquids. This puts an added burden on the starch-digesting enzymes in the small intestine (especially in the duodenum).

All carbohydrates (sugar, starch, vegetables, and fruits) must be thoroughly chewed. Carbohydrates require an alkaline environment for best enzymatic action. A good alkaline environment is the mouth. Only by chewing can we properly break down these foods. Furthermore, chewing is important for all foods in order to make them soft for swallowing and ultimate digestion.

Sometimes, however, chewing might be difficult. When our upper and lower teeth don't align properly, we have a malocclusion, which, along with over- or underbite, cross bite, loss of teeth, temporomandibular joint dysfunction (TMJ), muscle atrophy, or other problems of the teeth or jaws, can interfere with proper chewing. Anything that causes an impairment in chewing will interfere with the breakdown of cellulose (or other cell walls), and thus, with proper digestion.

The work of the salivary enzymes is short-lived for many reasons. First, food doesn't stay in your mouth very long before it is swallowed, and second, enzymes are sensitive to changes in pH (acidity and alkalinity). However, some enzymes can survive more acid (or more alkalinity) than others.

Our Vital pH Balance

Biochemically, every area of the digestive tract, every enzyme, and every food has an optimum pH somewhere between 0 and 14.

Excessive acidity and alkalinity can affect reaction rates and denature enzymes (just as high temperatures can). Acidity or alkalinity is measured on a fifteen-step scale (0 to 14) known as pH (potential hydrogen). A pH of 7 (water and milk are good examples) is considered neutral (neither acidic nor alkaline), while a high figure (such as 14 for lye) is considered alkaline and a low one (battery acid at 0) is acidic. The mouth has a usual pH range between 6 and 8 (near neutrality). Characteristically, each enzyme has an optimal pH range,

Table 2–2. The pH of Common Substances

Substance	pH
Concentrate lye	14
Ammonia	11
Baking soda	9
Pancreatic juice	8
Blood	8
Water	7
Milk	7
Urine	6
Coffee	5
Orange juice	4
Vinegar	3
Lemon juice	2
Gastric juice	2
Battery acid	0

which may be broad or narrow. At this optimum pH the enzymatic catalytic reaction occurs most rapidly.

Enzymes are referred to as having a certain pH rating. When an enzyme has a greater amount of hydrogen, it is considered an acid. If it has a lesser amount of hydrogen, it is called an alkali (or base). Therefore, a figure over 7 pH is alkaline (or base), while a figure under 7 pH is acidic.

Foods are also alkaline or acidic, though few foods are totally one or the other. Some are more acidic (or alkaline) than others. We usually decide whether a food is alkaline or acid by its taste. For example, lemon juice and vinegar are so acidic that we pucker our mouths. On the other hand, a food like milk is rather tasteless and smooth, so it is neutral.

As we age, forty, fifty, sixty years, our production of hydrochloric acid (among other things) in the digestive system slows down. This causes problems with enzymatic food digestion. For this reason, middle-aged individuals often have more difficulty digesting heavier meals than their younger counterparts.

Into the Stomach

After we swallow our food, it travels down the esophagus to the stomach. The stomach lies in the upper left region of your abdomen, just below the diaphragm. This pouch-like structure serves as a reservoir for food during the early stages of digestion. The trip to the stomach begins when a bolus (a mass of food) enters the pharynx (throat) and is grasped by the constrictor muscles, which move it onward into the esophagus. The esophagus (or gullet), lying behind the trachea (voice box), is a ten-inch long, collapsible, muscular tube.

A distensible sac, the stomach looks something like the letter J with the hook extending across the body's midsection with the final portion beneath the liver. The lining of your stomach is a prime source of glands, which supply hormones, enzymes, and hydrochloric acid essential for digestion. Specific hormones are released, which trigger the enzymatic breakdown of foods.

In the stomach, food is broken down into increasingly smaller particles and compounds by mechanical, chemical, and enzymatic means. Digestion in the stomach can continue for as long as an hour until food is combined with the secretions of the stomach.[4]

The stomach's acidic secretions contain the enzyme *pepsin* (gastric protease), which converts proteins into short chains of amino acids. Protein could not be used in your body without the help of pepsin. Among other things, proteins help give us strong muscles, resistance against disease, recovery from illness, healthy skin, strong skeletal structure, and a rich blood supply.

ENZYMES CONTAINED IN GASTRIC JUICES

Pepsin (gastric protease) causes protein digestion. The peptic and mucous cells secrete several different types of *pepsinogen*, which initially have no digestive activity. However, as soon as pepsinogen comes in contact with hydrochloric acid, it is immediately activated, forming *pepsin*. This is actually a safety factor, because if pepsin were secreted in its active form, it would digest the protein in the very cells that produced it. Pepsin is an active proteolytic enzyme with an optimal pH of 1.8 to 3.5. Pepsin becomes completely inactivated in a short time and has little proteolytic activity above a pH of about 5.[4] Hydrochloric acid is as necessary as pepsin for protein digestion in the stomach. So, if you don't have enough hydrochloric acid, the pepsinogen secreted cannot be turned into pepsin, and in turn, the proteins you eat will not be properly digested.

Hydrochloric acid is secreted by the parietal cells. Hydrochloric acid serves many functions; it denatures proteins, converts pepsinogens into pepsin, makes some minerals more absorbable (including calcium and iron), and helps kill bacteria. It also stimulates the production of hormones in the stomach, and promotes the uptake (into the bloodstream) of minerals and trace elements (some acting as cofactors).

Gastric lipase splits fat, which is then used to ward off allergic conditions and infectious viruses, and to cushion our bodies against traumatic blows and bruises. In addition, gastric lipases initiate the digestion of triglycerides.

Amylase enzymes in the stomach continue the work begun on carbohydrates by the salivary amylase.

Rennin is another important digestive enzyme. Rennin releases minerals from milk (calcium, iron, phosphorus, and potassium), which strengthen our nervous system, strengthen teeth and bones, stabilize the water balance, and so on. Rennin causes the coagulation of milk, changing the protein, casein, into a form we can use.

Mucus forms a coating on the stomach, helping it to resist the digestive forces of the gastric enzymes.

Intrinsic factor bonds with vitamin B_{12} and is necessary for intestinal absorption.

Other enzymes control gastric secretions.

To break down your food, the stomach produces between 1 and 2 liters of gastric juice daily. The most important of the stomach's jobs are:

1. To act as a reservoir, storing food until it can be transported to the duodenum (the first portion of the small intestine);
2. To mix food with gastric secretions, forming a semifluid mixture called *chyme*; and
3. To slow the emptying of food into the small intestine at a rate suitable for its proper digestion and absorption.

The Small Intestine: Where the Action Is

The movement of the stomach empties liquids into the duodenum of the small intestine continuously while food solids are reduced to the consistency of paste (now called *chyme*).

The small intestine is a coiled tube extending from the stomach to the large intestine. The greatest amount of digestion and absorption takes place in the small intestine, which is approximately 23 feet in length and 1 inch in diameter. The coiled intestinal loops are attached to the back abdominal wall by the fan-shaped mesentery (membrane).

The small intestine consists of three divisions: duodenum, jejunum, and ileum. Each division is a vital part of digestion.

The stomach gradually releases the *chyme* into the duodenum, which is shaped like the letter C. The duodenum forms a concavity in which the head of the pancreas lies. The duodenum (Latin for twelve at a time) owes its name to the fact that it is about twelve finger-breadths long. Substances are discharged into the duodenum at varying rates depending upon their caloric density and the effects of neuropeptides (or other hormones) that control gastrointestinal function.

When carbohydrates in the chyme enter the duodenum, they do so partly in the form of dextrins and maltose (produced by the action of salivary amylase), starches and glycogen (that escaped digestion in the mouth), or as sucrose and lactose (cane sugar and milk sugar).

Fats in the chyme enter the duodenum in the form of triglycerides (the most abundant of fats), as well as phospholipids, cholesterol, and cholesterol esters. Only a small portion of fat (that is, the short chain triglycerides) is digested in the stomach, as most fat digestion occurs in the small intestine.

Proteins in the chyme enter the duodenum mainly in the form of proteoses, peptones, and large polypeptides.

Hormones send signals to the gall bladder and pancreas requesting the enzymes necessary for *hydrolysis*. Hydrolysis is the process of digesting energy nutrients, the chemical reaction where water is absorbed by a large molecule, breaking it into smaller ones that are water-soluble, and thus, able to be used by the body's cells. You can see the importance of water in digestion; without it, enzymes can't do their work.

The pancreas, a large compound gland, is located parallel to and beneath the stomach and has many functions. In addition to insulin and glucagon production, the pancreas also secretes digestive enzymes for all major food types.

The glands of the small intestine produce intestinal juice (*succus entericus*). The presence of food and resulting distention of the intestine, along with stimulation of the vagus nerve, initiates flow of this juice.

Pendular movements of the intestine sweep forward, then backward, and consist of small constrictive waves. These movements mix the intestinal contents. At the same time, peristalsis moves the intestinal contents onward.

When you begin to eat, your food enters the stomach, greatly increasing the peristaltic activity of the small intestine. This *gastroenteric reflex* is why patients who have undergone intestinal surgery are not given food by mouth until their sutures have healed. This reflex could actually tear the stitches free.

Secretin causes the pancreas to produce a thin, watery juice that is rich in sodium bicarbonate, which helps to neutralize hydrochloric acid in the chyme. Because the duodenal lining is not as well equipped as the stomach to withstand the corrosive effect of gastric juice, neutralization of the hydrochloric acid is an important protective mechanism. People who develop a *duodenal ulcer* usually secrete too much hydrochloric acid, but this is only one of many factors that could be involved. Another important effect of sodium bicarbonate is that it produces a favorable duodenal pH (6.0 to 7.0) so that the pancreatic enzymes may operate efficiently.

Pancreozymin is a hormone that stimulates the pancreas to produce a thick, sticky secretion that is rich in amylase, lipase, and three proteases.

The body produces approximately 2.5 liters per day of pancreatic juices.[3]

Pancreatic amylase attacks glycogen, starches, and the majority of other carbohydrates (except cellulose), forming disaccharides and some trisaccharides. Amylase splits starch to maltose, which is then split (by *maltase*) to glucose. Sucrase splits sucrose to glucose and fructose, while lactase splits lactose to glucose and galactose. Amylolytic enzymes are able to degrade up to 300 grams of carbohydrate daily.[4]

Lipolytic enzymes (including lipase, phospholipase A_1 and A_2, and nonspecific esterase) split fats into glycerol and fatty acids. Although lipase can and does attack large fat globules, its action is much more effective when fats first are emulsified (split into tiny particles) by the action of bile salts from the gall bladder. Lipolytic enzymes can break down up to 175 grams of fat per hour.[4]

The *proteolytic enzymes* (including trypsin, chymotrypsin, elastases, and carboxypeptidases A and B) when synthesized in the pancreatic cells, are all in enzymatically inactive forms (*trypsinogen*, *chymotrypsinogen*, and *procarboxy-polypeptidase*). These enzymes only become activated after they are secreted into the intestinal tract. An enzyme called *enterokinase* (secreted by the intestinal lining when it comes in contact with chyme) activates trypsinogen (which can also be activated by already formed trypsin). Chymotrypsinogen is activated by trypsin to form chymotrypsin, and procarboxypolypeptidase is activated similarly. Proteolytic enzymes (proteases) are capable of breaking down up to 300 grams of protein per hour.[4]

Because activated trypsin (and other pancreatic proteolytic enzymes) would digest the pancreas itself, you can see why it is important that these juices not become active until secreted into the intestine. Fortunately, a substance called *trypsin inhibitor* is simultaneously secreted by the same cells that secrete the proteolytic enzymes. This substance prevents trypsin activation inside the secretory cells and pancreatic ducts. Further, trypsin inhibitor also prevents subsequent activation of all other pancreatic proteolytic enzymes since trypsin activates these enzymes, as well.

When a duct becomes blocked or when the pancreas is severely damaged, large amounts of pancreatic secretion become pooled in the damaged areas. The effect of trypsin inhibitor under these conditions is therefore overwhelmed, and the pancreatic secretions rapidly become activated, literally digesting the entire pancreas within a few hours. This causes a condition called *acute pancreatitis*. Often, this is lethal because of accompanying shock. However, even if it is not lethal, it still results in a lifetime of pancreatic insufficiency.

Trypsin, chymotrypsin, and carboxypolypeptidase are the more important of the proteolytic enzymes. By far the most abundant of these is trypsin. Chymotrypsin and trypsin split whole and partially digested proteins into various sized peptides (but do not cause the release of individual amino acids). Carboxypolypeptidase, on the other hand, splits the peptides into the individual amino acids, completing digestion of much of the proteins all the way to the amino acid state.

In addition to amylases, lipases, and proteases, the pancreas also produces *nucleases* (such as deoxyribonuclease and ribonuclease) and other enzymes, such as procolipase and trypsin inhibitor.

Table 2-3 shows the digestive enzymes from the pancreas and their precursors.

The Jejunum and the Ileum

After leaving the duodenum, the chyme now enters the jejunum and ileum of the small intestine where most digestion and absorption

Table 2-3. Pancreatic Digestive Enzymes and Their Precursors

INACTIVE PRECURSOR	ACTIVE ENZYME
Proteolytic Enzymes	
Trypsinogen	Trypsin
Chymotrypsinogen	Chymotrypsin
Proelastase	Elastase
Procarboxypeptidases A and B	Carboxypeptidases A and B

take place. Absorption in the small intestine is a lot like sorting products on an assembly line. As they pass by, individual nutrients are selected, pulled out, and transported into the blood stream through the walls of the small intestine.

A disturbance in intestinal wall absorption can result in the entire digestive system shutting down. This type of disturbance can also cause gastrointestinal distress.

Enzymes play a decisive role in food absorption from the gut to the bloodstream and are essential in the transport of nutrients.

The absorption of sugars, peptides, amino acids, and fatty acids proceeds in the jejunum, while bile salts and vitamin B_{12} are absorbed in the ileum. By the time food residues pass the ileum, most usable food molecules have been digested and absorbed, so that only small amounts of carbohydrates, other macronutrients, and fiber are delivered to the colon.

Enzymes are considered to be macromolecules because of their large molecular size. For years, it was believed that the body could not absorb intact macromolecules, including enzymes. It was felt they had to first be broken down into smaller parts (amino acids) in order to be absorbed by the small intestine into the bloodstream. Modern technology has now made it possible to accurately study the absorption of enzymes (and other macromolecules) from the gut into the blood and/or lymphatics. Although there are several methods available to study this absorption, immunofluorescent techniques and radio chromatography appear most reliable.

From these studies, we have learned that a frequent mechanism for enzyme absorption is the *pinocytotic transfer by the cells of the intestinal wall.* Pinocystosis is the process by which whole molecules are engulfed and absorbed. After connection to a receptor (in the intestinal wall), the enzymes are absorbed into this wall, guided through the intestinal cells in vesicles (small sacs), and released into the blood.

The Large Intestine: Getting the Waste Out

Finally, the large intestine completes the job of digestion. The large intestine extends from the end of the ileum to the anus and is about 4½ to 5 feet in length and 2½ inches in diameter. It is segmented and larger in caliber than the small intestine.

The colon is subdivided into ascending, transverse, descending and sigmoid segments. The principal functions of the large intestine are to concentrate, store, and excrete food wastes. The colon contains large numbers of bacteria, which produce enzymes that act on the remaining food residues, fiber, cells, and mucus discarded from the upper intestinal tract. A certain amount of subtle utilization of these waste products can take place in the large intestine and rectum. The entire gut (particularly the large intestine and rectum) is alive with microorganisms which seek out additional nutrients. The remaining unusable bulk travels from the small to the large intestine, where water is absorbed and waste excreted from the rectum as stool.

One would think that parasites (foreign organisms) located in our bowels would be attacked and eliminated by our immune system, but they're not. Instead, foreign bacteria in the gut feed on these parasites. These bacteria, then, produce by-products, which our bodies use. An example is vitamin K (necessary for blood coagulation).

Digestive Helper Organ

Of all the organs in the human body, the liver is the largest and the busiest. The liver produces bile, which is considered to be both an excretion and a secretion. Formed continuously by the hepatic cells (average daily volume is about 600 to 800 milliliters), this greenish-yellow fluid contains water, bile salts, bilirubin, cholesterol, fatty acids, lecithin, and various inorganic salts. Only the *bile salts* (formed from cholesterol) have important activities in digestion, and two major actions. First, bile salts help fatty acid absorption from the intestinal tract. Second, they emulsify fats by lowering the surface tension, allowing large fat globules to be broken into tiny particles. This is a valuable function because the fat-splitting enzymes can attack many small units more easily and efficiently than one large unit.

The liver excretes *bilirubin*, or bile pigment. Bilirubin, tied to a plasma protein, travels to the liver in the bloodstream, where it is made water-soluble. From the liver, it is excreted into the bile.

The liver plays its most vital role in protein metabolism. For example, the hepatic cells form essentially all of the plasma proteins, except for the gamma globulins. In carbohydrate metabolism, the liver helps to maintain a normal blood-glucose concentration. Although fat (lipid) metabolism probably can take place in most cells of the body, certain activities occur much more rapidly in the liver. Specific functions would include the synthesis of fat from glucose and amino acids, plus the formation of cholesterol, phospholipids, and lipoproteins.

Table 2-4. Secretions of Intestinal Juices

SECRETION	VOLUME SECRETED DAILY
Saliva	1000 ml
Gastric secretion	1500 ml
Pancreatic secretion	1000 ml
Bile	1000 ml
Small intestine secretion	1800 ml
Large intestine secretion	200 ml

Jump-Start Your Digestion

How can some people eat nutritious foods and yet be continually tired, develop chronic diseases, and/or age prematurely? Quite possibly, it could be poor digestion and/or absorption of foods. In other words, an individual could be eating a healthy diet, but the nutrients aren't getting to the body. Literally, one can eat the best and yet

(*text continues on page 30*)

Table 2–5. Digestive Enzymatic Secretions

SOURCE	ENZYME	pH	ACTION
Salivary glands	Salivary amylase	6.0 – 7.0	Salivary glands secrete saliva in response to the presence of food in the mouth. Salivary amylase breaks down carbohydrates.
Stomach glands	Pepsin	1.0 – 2.0	Hydrochloric acid in the stomach converts pepsinogen to pepsin. Pepsin breaks down proteins.
	Rennin	4.0	Rennin breaks down casein in milk products.
	Mucus		Mucus protects the stomach from hydrochloric acid.
	Gastric lipase		Lipase breaks down fat and initiates the digestion of triglycerides.
	Amylase		Amylase breaks down carbohydrates.
	Intrinsic factor		Bonds with B_{12} and is necessary for intestinal absorption.
Pancreas	Trypsin	5.2–6.0	Enterokinase in the small intestine activates trypsinogen to form

SOURCE	ENZYME	pH	ACTION
			trypsin, which breaks down protein peptides.
	Chymotrypsin	8.0	Is converted from chymotrypsinogen and breaks down protein peptides.
	Elastase		Converted by trypsin from proelastaste to elastase. Breaks down peptides.
	Carboxypeptidase		Activated by trypsin (from procarboxypeptidase). Breaks down polypeptides.
	Amylase	7.1	Breaks down starch and glycogen.
	Lipase	8.0	Breaks down fats.
	Ribonuclease		Breaks down ribonucleic acid.
	Deoxyribonuclease		Breaks down deoxyribonucleic acids.
	Cholesteryl ester hydrolase		Breaks down cholesteryl esters.
	Phospholipase A		Breaks down phospholipids.
Small intestine	Amino peptidase		Breaks down polypeptides.
	Dipeptidases		Breaks down dipeptides.
	Sucrase	5.0–7.0	Breaks down sucrose.
	Maltase	5.8–6.2	Breaks down maltose.

(*continued*)

Table 2–5, continued

SOURCE	ENZYME	pH	ACTION
	Lactase	5.4–6.0	Breaks down lactose.
	Phosphatase	8.6	Breaks down organic phosphates.
	Isomaltase		Breaks down glucosides.
	Polynucleotidase		Breaks down nucleic acid.
	Nucleosidases		Breaks down purine or pyrimidine nucleosides.
	Trehalase		Breaks down trehalose.

the body is starving. There is a road block on the conveyor belt of life! You're not what you eat, but what you absorb, says Jeff Bland.

As you can see, enzymes are fundamental to the digestive process. But, what are some of the problems that might interfere with this process? Chapter 3 discusses some conditions and disease states, which inhibit our ability to properly digest food.

References

1. David Dressler and Huntington Potter, *Discovering Enzymes*. New York: Scientific American Library, 1991, 201.
2. Eleanor Whitney and Eva Hamilton, *Understanding Nutrition*, 4th Ed. Minneapolis, Minn.: West Publishing Co., 1987, 132.
3. *Best and Taylor's Physiological Basis of Medical Practice*, John B. West, Editor. 12th ed. Baltimore: Williams and Wilkins, 1990, 648.
4. Arthur G. Guyton, *Textbook of Medical Physiology*. Philadelphia: W. B. Saunders Co., 1991, 700–727.

Chapter 3

Enemies of Digestion

Now that we know why it's important to eat "live" food and how our bodies break down, digest, and absorb foodstuffs in order to use them as fuel, everything is OK, right? Wrong! Sometimes, eating properly is not enough. Certain individuals have genetic or congenital problems, which interfere with normal digestion and absorption of nutrients. Usually the reason for a disorder is unknown, though modern science has made enormous advances in the last few years in learning to treat these conditions. Problems that interfere with digestion and absorption can be mechanical (a difficulty in properly chewing food), biochemical (a genetic predisposition or lack of an enzyme), or stress (of any kind). Any of these problems can interfere with proper digestion, leading to illness and in some cases, death.

In addition, anything that alters the secretion of hydrochloric acid in the stomach will affect the secretion of enzymes. Conditions such as accidental trauma, chemotherapy, chronic and terminal illness, chronic pain, emotional stress, mental depression, physical trauma, and surgery can cause changes in hydrochloric acid secretion. This, then, would disrupt the flow and activation of enzymes and, ultimately, affect the absorption of food.

Mechanical Problems

As previously mentioned, most fruits and vegetables are covered with an indigestible cellulose layer. If we don't properly chew our food, the membrane around the food can't be adequately broken, which will interfere with digestion and absorption. Unlike many animals, the human body does not produce the enzyme *cellulase*, which breaks down cellulose. Cellulose, by the way, is another word for fiber. Is there anyone alive who doesn't know the value of fiber in our diets? Fiber is important because it adds bulk—which, in turn, improves elimination. But, don't we want more than just bulk from our fruits and vegetables? Don't we also want to get the benefit of all nutrients, including any vitamins, minerals, and enzymes these foods might offer? To do this, we must thoroughly chew our food and break that cellulose membrane.

But what happens if your teeth aren't properly aligned, interfering with your ability to adequately chew foods? What if you have a sore tooth on one side and only chew on the other side? What if you are missing teeth? How about misaligned jaws? All of these conditions can interfere with your ability to properly chew your foods. Remember, digestion begins in the mouth, so do all you can to give it a good start.

Does it hurt to chew? Do you have false or loose teeth? Do your teeth not approximate (come together) properly?

Your ability to chew is important for proper digestion. Chewing is negatively affected by pain and dysfunction, which can have at least three reasons for their origin: genetic and congenital factors, developmental factors, or traumatic factors, as seen in Table 3–1.

Under psychological stress or nervous tension, we might tend to clench our teeth. The degree of clenching (or bruxing) increases and stresses the muscles, causing repeated grinding of the teeth and wearing away of tooth surfaces.

Whatever the reason, the bottom line is, if you can't chew or if chewing is painful, you either won't chew, you'll chew less vigorously, or you'll take more liquid and/or mashed foods so that chewing is not necessary. As a result, your mouth does not have a chance to do its job. The mouth's food mixing and grinding functions are reduced or shut down. This means the salivary enzymes can't do their digestive jobs in beginning to break down carbohydrates.

Further, digestive or malabsorption problems can often become apparent with conditions such as gingivitis (inflammation of the gums) or increased tooth decay.

Salivary production has a definite effect on digestion. We're all familiar with Dr. Pavlov and his studies on salivary production in animals. What Pavlov found was that the quantity and quality of saliva varied according to what kind of food he offered the dogs as

Table 3–1. Factors Affecting Chewing and Their Possible Causes

CONDITION	GENETIC OR CONGENITAL PREDISPOSITION	DEVELOPMENTAL	TRAUMATIC
Cleft lip	X		
Cleft palate	X		
Overbite	X	X	X
Underbite	X	X	X
Cross bite	X	X	X
Temporomandibular Joint Dysfunction (TMJ)	X	X	X
Weak or absent jaw muscle—movement	X	X	X
Ligament imbalance	X	X	X
Calcified tissues	X	X	X
Psychological stress or disorders	X	X	X

a stimulus. For example, meat powder produced more saliva flow than raw meat. On the other hand, sand was a more potent stimulus than smooth stones in causing saliva to be secreted. In addition, when he fed the dogs foods high in acids, they produced saliva rich in proteins (which acted as buffers, neutralizing the excessive acidity). He found that dry substances (such as meat powder and sand) produced an abundant flow of saliva, which was probably the body's attempt to compensate for the dryness of the material.

Lack of normal saliva production can have serious, even deadly, effects. For example, removal of the salivary glands in the newborn rat disturbs the swallowing to such an extent that the animal can starve to death. Further, if both parotid glands of a horse are obstructed, the animal can't chew or swallow. The horse will then eat less (and more slowly), which will lead to serious illness, and possibly, slow death.

Biochemical Problems

Sometimes eating right and chewing properly aren't enough. The human body, like any complex machine, can malfunction. Perhaps your body simply doesn't produce the enzyme necessary to break down various components of your food. For instance, if you lack the enzyme necessary to break down the amino acid *phenylalanine*, you have a condition called phenylketonuria, caused by the defective enzyme *phenylalanine hydroxlyase*. The clinical features of this disease include neurologic symptoms and mental retardation. The only treatment is to control phenylalanine intake. (Note: Aspartame, a sweetener widely used in diet sodas and other foods, is made of two amino acids: aspartic acid and phenylalanine, so should be avoided by those who should limit their phenylalanine intake.)

A multitude of biochemical problems related to enzyme production or utilization can occur in the human body. Some conditions interfere with the activation of digestive enzymes, such as hypochlorhydria and achlorhydria.

Hypochlorhydria and Achlorhydria

Hypochlorhydria refers to diminished acid secretion—not enough hydrochloric acid is produced. Usually, when acid is not secreted, neither is pepsin. A lack of either or both will interfere with digestion. Even when pepsin is secreted, a lack of sufficient acid would prevent it from functioning because pepsin requires an acid medium for activity.

Achlorhydria simply means that the stomach secretes *no* hydrochloric acid. This condition is diagnosed by measuring the pH of the gastric secretions.

Although achlorhydria is associated with depressed and even a lack of stomach digestive capability, the overall food digestion throughout the entire gastrointestinal tract may still be almost normal. This is because trypsin and other enzymes secreted by the pancreas are capable of digesting virtually all of the dietary protein. But do you want "almost normal" digestion—or do you want optimal digestion? In addition, did you know that calcium cannot be properly absorbed without adequate amounts of hydrochloric acid? Consider this before you take another antacid.

Some physicians treat achlorhydria and hypochlorhydria with hydrochloric acid, which enables the digestive enzymes to be fully activated, leading to better digestion.

It's interesting to note that arthritics often suffer from these conditions. In one investigation, researchers studied seventy patients, half with rheumatoid arthritis, half with osteoarthritis. What they found was that most people suffering from arthritis don't have enough stomach acid! They found a dramatic incidence of achlorhydria in their rheumatoid arthritis patients (28.6 percent) and their patients suffering from osteoarthritis (25.7 percent) as compared to normal subjects in the same age group (10.8 percent).[1] We'll talk more about arthritis in a later chapter.

Pancreatic Insufficiency

For a variety of reasons (including pancreatitis, surgical removal of the pancreas, a blocked pancreatic duct, or insufficient output), the pancreas may not secrete its juice into the small intestine. This condition leads to a loss of digestive enzymes (including carboxypolypeptidase, chymotrypsin, pancreatic lipase and amylase, trypsin, plus others), resulting in improper digestion. Without digestive enzymes, one-third to one-half of the carbohydrates and proteins, and the majority of fats entering the small intestine will not be absorbed. Because large amounts of food cannot be digested, this condition results in loss of nutrition. In addition, the failure to properly break down ingested nutrients leads to excretion of large amounts of fatty feces (steatorrhea).

Cystic Fibrosis

Steatorrhea is a common complaint among those suffering from cystic fibrosis (CF). CF is an inherited disease, which primarily affects the gastrointestinal and respiratory tracts. It typically is characterized by chronic obstructive pulmonary disease, pancreatic insufficiency (in 85 percent of CF patients),[2] and abnormally high loss of electrolytes through perspiration. CF patients are frequently given enzymes to treat the pancreatic insufficiency, thus improving digestion of fat, which, in turn, decreases the steatorrhea.

Several studies have been conducted to test the efficacy of pancreatic enzyme products in treating the steatorrhea in CF patients. One study at the University of Wisconsin compared an enterically coated enzyme product with a noncoated preparation, as well as a placebo. (Enterically coated products are designed to dissolve in the small intestine, rather than the stomach.) What they discovered was that CF patients experienced a significant improvement in both nitrogen and fat absorption when receiving either enzyme preparation as opposed to the placebo. In addition, the enteric-coated product improved fat absorption as compared with the conventional enzyme capsule.[3]

Enzyme therapy has been taken one step further in the case of cystic fibrosis. CF is characterized by a thick buildup of mucus in the lungs and gastrointestinal tract. The presence of this mucus in the body hinders digestion and therefore, promotes infection. Traditional treatments of CF include antibiotics and bronchodilators. Chest percussion is also employed on a regular basis, to help break up the mucus.

In the past, children with cystic fibrosis were only expected to live to age five. Now, life expectancy is up near age thirty, due, in part, according to the Cystic Fibrosis Foundation, to innovative new treatments.

One such innovative treatment is Pulmozyme (dornase alfa), launched in 1994. This enzyme (a purified solution of recombinant human deoxyribonuclease) is administered in an aerosol mist. It helps the CF patients to breathe easier because it breaks down the accumulated waste matter in their lungs. According to Genentech, Inc., the makers of Pulmozyme, it significantly reduces the risk of serious respiratory tract infections and also improves lung function.

Pancreatitis

But CF patients aren't the only ones with pancreatic insufficiency. This problem can occur for other reasons, including pancreatitis. Pancreatitis means "inflammation of the pancreas." Over 80 percent of hospital admissions for acute pancreatitis are due to biliary tract disease and alcoholism. The rest (20 percent) are caused by hereditary and other factors.[4] If the inflammation resolves when the cause is removed, we call this acute pancreatitis. The chronic form (usually alcohol-related) means that tissue destruction is irreversible.

It is believed that drinking more than 100 grams of alcohol per day for several years may cause obstruction of pancreatic ducts. Over time, this obstruction becomes more widespread and causes additional abnormalities. After three or five years, the first clinical episode of pancreatitis may occur, presumably because of premature activation of pancreatic enzymes.

Acute Pancreatitis

The symptoms of acute pancreatitis may be severe. Almost all patients with this condition suffer severe abdominal pain that radiates through to the back. It's important for us to be aware that the cause of this pain is not due to a back injury, but rather to a potentially life-threatening pancreatic disorder. If you think you have this condition, see your physician immediately. In spite of improved intensive care, because of the potential for shock, the possibility of death is still very high. Mortality is 20 to 80 percent depending on the severity of the attack.[4]

Chronic Pancreatitis

The most common cause of chronic pancreatitis in the United States today is alcoholism. Only rarely does the chronic stage of this disease develop from other factors. Like acute pancreatitis, the chronic form may be very painful; however, often the patient is pain-free. The patient may lose weight, have trouble tolerating fatty foods, and may suffer from steatorrhea and/or diabetes. Chronic pancreatitis leads to pancreatic insufficiency. Eventually, the cells in the pancreas (that secrete digestive enzymes) are destroyed, and abdominal pain may actually subside. In time, pancreatic secretions will reduce and the patient will develop steatorrhea. Destruction of the islet cells of the pancreas will result in reduced insulin secretion and glucose intolerance.

Therapy is limited to dietary measures, antacid administration, and high levels of pancreatic enzymes. Surgical intervention is usually only considered when the cause of the disease is clear and if it is indicated.

Enzymes have been used therapeutically in the treatment of pancreatic insufficiency for some time. Unfortunately, however, stomach acid can inactivate the lipase, reducing the effectiveness of the enzyme products—meaning that large numbers of tablets are required to provide adequate supplementation. Researchers have investigated ways to make the enzymes impervious to stomach acid. One way is by coating them with a substance that is resistant to stomach acid and will dissolve only in the small intestine. One study analyzed the effects of antacids, cimetidine, and enterically coated enzymes in the treatment of alcoholics suffering from pancreatitis. The researchers found that steatorrhea was significantly reduced by all treatment methods, but especially when cimetidine (a potent inhibitor of gastric acid secretion) and the enzymes were given concurrently. In fact, this combination abolished steatorrhea in four of the six patients studied.[5]

Another study compared enzyme tablets containing 3,600 units of lipase with a microencapsulated enteric preparation containing 2,005 units of lipase per capsule. Results indicated that both preparations significantly reduced fecal fat and nitrogen excretion, as opposed to the use of no enzyme at all.[6]

Acid-resistant lipase from *Aspergillus* has been used successfully to treat pancreatic insufficiency in dogs. In fact, the researchers found that a very small amount of acid-stable lipase (400 milligrams) was as effective as 10 milligrams of pancreatin in treating steatorrhea.[7]

It has been known for some time that patients with pancreatic insufficiency do not properly absorb fat and protein. For a long time, however, it was assumed that they *did* absorb starch. We now know that this, too, is adversely affected. Researchers measured the absorption of bread (containing 100 grams of rice flour) in five patients with pancreatic insufficiency. This amount of carbohydrate (which can be completely absorbed by normal subjects) was malabsorbed in each of the five patients. In fact, they all showed large increases in breath-hydrogen concentrations, showing that starch malabsorption might play a role in the diarrhea and malnutrition found in patients with pancreatic insufficiency.[8]

The production and release of pancreatic enzymes is affected by many factors. For example, a great strain is placed on the pancreas from excessive intake of refined carbohydrates and alcohol. Such poor eating habits can decrease the amount of enzymes produced by the pancreas and contribute to adult-onset diabetes, food allergies, and pancreatitis.

Enzyme substitution therapy seems to have definite value in pancreatitis.

Celiac Disease

Celiac disease is an immunologic reaction to the gluten in wheat, rye, or oats. It is the gluten (the elastic protein) in grains that gives dough its structure and cohesiveness. It is generally agreed that gliadin (a component of gluten) is the toxin that causes celiac disease.[9] Unfortunately, when patients with this disorder eat anything containing gluten, the cells that line the small intestine waste away. This causes malnutrition, anorexia, vomiting, diarrhea, steatorrhea, stunting of growth, and anemia. Completely removing gluten from the diet will eliminate these symptoms. It will also restore the integrity of the intestinal lining, but some immunologic abnormalities may remain. We don't know why some people can't tolerate gluten, but researchers are making advances.

Almost 1 in every 2,500 people in the United States suffers from celiac disease. But, the incidence in southwest Ireland is much more dramatic, where 1 of every 300 people suffers from the disease.[10]

In one study in Ireland, researchers removed carbohydrates from the gliadin, making the proteins nontoxic to celiac patients. They did this by treating it with the enzyme *carbohydrase*, from *Aspergillus niger*. They then took this treated gliadin, prepared and baked it in gluten-free bread. Gliadin treated in this manner did *not* damage the intestinal lining of the celiac patients.[9] Yet another example of what enzymes can do for you.

Papain has also been used to treat celiac disease. Crude papain (the dried and powdered latex of the *Carica papaya*) contains (among other enzymes) papain, chymopapain, peptidase, and glutamine cyclotransferase. Papain is used as a digestive aid and as a meat tenderizer. One study examined the use of papain in treating a patient suffering from celiac disease. After six months on a gluten-free diet, the patient still exhibited symptoms of the disease (including steatorrhea). He was then given papain tablets with every meal. After four weeks, his absorption normalized. At this point, despite abandoning his gluten-free diet, he remained symptom-free.[11]

Lactose Intolerance

Can you drink milk? Or do you, like some people, experience abdominal pain and diarrhea if you drink milk or eat anything that contains milk, like ice cream? If so, you may have lactose intolerance. An insufficiency of *lactase*, the enzyme responsible for breaking down lactose (milk sugar) in the small intestine, can cause this condition. If lactose cannot be properly broken down, it will sit there, undigested, in the colon, where it undergoes bacterial action, ferments, leading to diarrhea and pain.

Lactose intolerance is found more frequently in certain cultures. Genetic absence of lactase beyond the age of five or six occurs at remarkably high proportions among Asians, Africans, Native Americans, and African-Americans, as well as among a significant percentage of healthy Caucasians.

Table 3–2. Incidence of Lactose Intolerance[12]

Asians	85 to 95 percent
Africans	50 to 99 percent
African-Americans	70 to 75 percent
Native Americans	85 to 95 percent

Lactose intolerance is also called lactase deficiency, lactose malabsorption, and milk intolerance. This condition can be caused by: 1) congenital absence of lactase; 2) a decrease in the level of lactase; or 3) damage to the intestinal lining by viruses, bacteria, gluten, and so forth. Lactose intolerance, however, is not an inevitable consequence of lactase deficiency. Many people can consume modest amounts of lactose-containing foods with little difficulty. Some people can consume milk and milk products if they add the enzyme, lactase, or use fermented products such as cheese or yogurt.

How do we know that adding lactase will do the trick? One study analyzed eighty-nine children (average age eleven years) who suffered from lactose intolerance to see if tablets containing lactase (also called beta-galactosidase) would have any effect on their symptoms. Before drinking a lactose-containing solution, the subjects were given either a tablet containing lactase or a placebo. The breath of the children was analyzed for hydrogen content every thirty minutes for two hours. Hydrogen levels in the breath are a good measure of the body's ability (or inability) to break down lactose. Those patients given the placebo showed an increase in hydrogen production, along with abdominal pain (in 89 percent), bloating (83 percent), diarrhea (61 percent), and flatulence (44 percent). Those subjects given the lactase-containing tablets *before* they were given the lactose showed a *reduction* in both hydrogen excretion and abdominal symptoms.[13]

Another study showed that adding enzymes to milk five minutes before consumption produced a 62 percent reduction in breath hydrogen excretion. This also significantly reduced the symptoms. According to the researchers, much of the world's population suffers from primary lactase deficiency—a deficiency that can be corrected with the addition of enzymes.[14]

Enzymes from *Aspergillus oryzae* have also been used effectively to treat lactose intolerance. Researchers at the Institute of Nutrition of Central America and Panama, in their study on lactose maldigestion, found that adding *Aspergillus oryzae* to milk at the time of ingestion reduced hydrogen excretion.[15] As mentioned above, the hydrogen breath test is an effective way to measure incomplete carbohydrate absorption.

Lactase-containing products are available at most health food stores in liquid and capsules, and in some places it is possible to purchase lactase-treated milk.

Stress

Did you know that stress of any kind can interfere with proper digestion and absorption of your foods? This includes emotional stress, as well as stress from injury, surgery, and toxins.

When your brain perceives stress, it kicks a series of responses into gear. Because the brain's reaction is to fight or take flight, it realizes that digestion is not a priority. The glands and organs involved in digestion shut down as the body diverts energy to handle the increased needs of respiration, circulation, and locomotion. Less blood flows to the GI tract and digestion virtually stops.

Though an occasional stressful situation probably causes no long-term detrimental effects to our digestion, what happens when you are constantly under stress, for whatever reason? One condition which develops from our response to stress is irritable bowel syndrome. This condition is marked by pain, abdominal distension, and alteration in bowel habits and is believed to be due to an inappropriate response, by the intestinal wall, to stress.

If we are constantly in a state of stress, our digestion will suffer. We will not be able to adequately absorb the nutrients in our food, so our nutritional status will diminish.

To Improve Your Digestion

Eat more enzymes! The quantity and type of enzymes used in digestion depend on the type and state of the food you eat. Some foods can actually aid your digestion. For instance, fresh, naturally ripened pineapples are rich in the protein-splitting enzyme, bromelain. However, because enzymes are very heat sensitive, canned pineapples have only a minimum amount of bromelain.

Heating food can destroy virtually all the active enzymes. We would have better digestion if we started each meal with a fresh vegetable salad, if we steamed our vegetables, and used salt sparingly (an indirect enzyme inhibitor). When eating foods difficult to digest, such as meat, eat plenty of enzyme-rich foods (such as raw sauerkraut, fresh herbs, garlic and onions).

According to Dr. Edward Howell (in his book, *Enzyme Nutrition, The Food Enzyme Concept*), a diet without enzymes can result in enlargement of the pancreas. He believes that eating fresh or raw foods (rich in enzymes) can inhibit the pancreas, preventing its enlargement. He says an enlarged pancreas is correlated to an increase in chronic degenerative diseases and cancer.

In some of their eating habits, the Japanese have an instinctive feel for healthy nutrition. They eat raw fish and raw seafood (hopefully from unpolluted waters). They use relatively large quantities of soy sauce (high in enzymatic activity) and cook their vegetables at a minimum. Unfortunately, they also have a high intake of salt. The leading cause of death and disability in Japan is hypertension, which has been linked to high sodium intake.[16]

Soy sauce is probably the world's oldest enzyme agent, and has been recognized in Asia for centuries for its digestion-promoting properties. Soy sauce is made by mixing soy flour with barley or rice, then fermenting this mixture with the help of *Aspergillus oryzae* (a fungus which contains very active enzymes called pronases). These pronases break down meat protein.

Americans, on the other hand, frequently don't use common sense when eating. Many of us don't eat the right foods, correctly prepared, at the best times, or in proper quantities. This can result in digestive disturbances, deposition of fat, and obesity, and can be associated with additional health problems, such as cardiovascular disorders. In addition, we eat on the run, which increases our already too-high stress levels, in turn affecting our digestion.

One way to support our overworked digestive system is to take natural digestive enzymes. Pepsin is probably the best known and is essential for protein digestion. Enzyme preparations contain many enzymes capable of breaking down proteins, fats, and carbohydrates. Some sources of digestive enzymes include pineapple, papaya, malt, *Aspergillus niger*, *Aspergillus oryzae*, ox bile, and hog pancreas. Enzymes are available in capsules, tablets (chewable, too), liquids, sublingual, powders, and granules. Because most digestion takes place in the small intestine, some digestive aids are enterically coated to dissolve there, and not in the stomach. As we age, the quantity of our body's enzymes decreases and so does the quality. The speed with which this happens is greatly influenced by our lifestyle and diet. An enzyme-poor diet can overtax an already deficient system.

Today's current scientific research indicates that large enzyme molecules can be absorbed from the gut, passing into the circulatory and lymphatic systems and, ultimately, to every cell of the body.

Enzymes are large protein molecules. Recent research by Professor G. Gebert of the Department of Physiology at the University of Ulm, Germany, indicates that we do, in fact, absorb whole enzymes into the bloodstream by *pinocytosis*.[17]

A person suffering from an enzyme deficiency can swallow the needed helpers by taking digestive enzyme supplements. In addition, since many chronic disorders involve disturbed enzyme function, it seems logical to take supplemental enzymes.

Digestion is a vital process in sustaining our body's health status and affecting our rate of aging. It is influenced by: 1) the time, quantity, and quality of food intake; 2) the capability of the digestive system to execute its various functions; 3) the capacity of the body to absorb the needed nutrients; and 4) the body's ability to utilize the products, once absorbed. We have seen the essential part enzymes play in breaking down our basic foods, and the need for digestive aids when our bodies' own mechanisms lose their effectiveness. Finally, enzymes (when taken orally) can be absorbed from the small intestine into the bloodstream and lymph, and then into the body.

IN REVIEW

Enzymes can help digestive problems, such as the following:

Hypochlorhydria
Achlorhydria
Pancreatic insufficiency
Acute pancreatitis
Chronic pancreatitis
Cystic fibrosis
Celiac disease
Lactose intolerance
Intestinal toxemia
Malabsorption
Steatorrhea
Food allergies

References

1. E. W. Hartung and O. Steinbrocker, "Gastric Acidity in Chronic Arthritis," *Annals of Internal Medicine* 9:252–257 (1935).
2. P. A. DiSant'Agnese, "Fibrocystic Disease of the Pancreas with Normal or Partial Pancreatic Function," *Pediatrics* 15:683–697 (1955).
3. Elaine H. Mischler, Sara Parrell, Philip M. Farrell, and Gerard B. Odell, "Comparison of Effectiveness of Pancreatic Enzyme Preparations in Cystic Fibrosis," *American Journal of Diseases of Children* 136:1060–1063 (1982).

4. Robert Berkow, Editor, *The Merck Manual.* Rahway, New Jersey: Merck Sharp & Dohme Research Laboratories, 1987, 763–764.
5. Patrick T. Regan, Juan R. Malagelada, et al., "Comparative Effects of Antacids, Cimetidine, and Enteric-Coating on the Therapeutic Response to Oral Enzymes in Severe Pancreatic Insufficiency," *New England Journal of Medicine* 297:854–858 (1977).
6. David Y. Graham, "An Enteric-Coated Pancreatic Enzyme Preparation that Works," *Digestive Diseases and Sciences* 24(12):906–909 (1979).
7. S. M. Griffin, D. Alderson, and J. R. Farndon, "Acid Resistant Lipase as Replacement Therapy in Chronic Pancreatic Exocrine Insufficiency: A Study in Dogs," *Gut* 30:1012–1015 (1989).
8. Robert D. Mackie, Allen S. Levine, and Michael D. Levitt, "Malabsorption of Starch in Pancreatic Insufficiency," *Gastroenterology* 80(5):1220 (1981).
9. J. J. Phelan, et al., "Celiac Disease," *Clinical Sciences and Molecular Medicine* 53:35–43 (1977).
10. Gluten Intolerance Group, Fact Sheet on Celiac Sprue. Seattle, Wash., 1982.
11. M. Messer and P. E. Baume, "Oral Papain in Gluten Intolerance," *Lancet* 2(S):1022 (1976).
12. *The Surgeon General's Report on Nutrition and Health 1988.* Washington, D.C.: U.S. Department of Health and Human Services, 1988, 417.
13. M. S. Medow, et al., "Beta-Galactosidase Tablets in the Treatment of Lactose Intolerance in Pediatrics," *American Journal of Diseases of Children* 1261 (Nov. 1990).
14. Jorge L. Rosado, Noel W. Solomons, Ruben Lisker, and Hector Bourges, "Enzyme Replacement Therapy for Primary Adult Lactase Deficiency," *Gastroenterology* 87:1072–82 (1984).
15. Caroline Barillas and Noel W. Solomons, "Effective Reduction of Lactose Maldigestion in Preschool Children by Direct Addition of ß-Galactosidases to Milk at Mealtime," *Pediatrics* 79:766–772 (1987).
16. Jane Brody, *Jane Brody's Nutrition Book.* Toronto: Bantam Books, 1987, 200.
17. G. Gebert, *Physiologie.* Stuttgart, Germany: Schattauer Publishing Co., 1987.

CHAPTER 4

Toxic Effects

YOU'RE EATING complex carbohydrates, plenty of enzyme-rich fruits and vegetables, and taking enzyme supplements. You're doing all you can to eat right. But are you really doing everything you can? Is your air, food, or water contaminated with pesticides, heavy metals, or other toxins? If so, you may not be getting the nutrients and the enzymes you think you're getting, or your body might not be processing them in the best way possible.

PESTICIDES

Dizziness, diarrhea, headaches, tremors, vomiting, or weakness can all be neurotoxic effects of pesticides. Pesticides are everywhere: in the air we breathe, the water we drink, in and on the food we eat. They are serious hazards to our health.

The U.S. government says a pesticide (which includes insecticides, herbicides, fungicides, rodenticides, nematocides, etc.) is any substance or mixture intended to prevent, destroy, repel, or mitigate any insect, rodent, nematode, fungus, or weed or any other form of life declared to be a pest.[1]

Carbamate and organophosphorous insecticides are the most common causes of agricultural poisoning in the United States and are also the most neurotoxic. They pose a significant threat to every American and especially to those four to five million who work in agriculture.

Although pesticides were initially developed by the Germans for use as nerve gas in World War II, they are now used to control and eliminate pests. Our use of pesticides since their discovery has exploded. In fact, our use is now thirty-three times greater than it was in 1945![2]

Pesticide use can produce aesthetic, health, and economic benefits. Aesthetically, by controlling pest damage, pesticides help us to produce a better-looking product. You are more likely to buy a shiny red apple than a dull, insect-infested one. Some crops are highly dependent on this type of pesticide use. For example, it is estimated that 60 to 80 percent of pesticide application to oranges is related to reducing cosmetic damage.[3] A lot of tomatoes grown in California go into tomato sauce and other crushed products. But did you know that almost two-thirds of the pesticides applied to these tomatoes is put there solely for cosmetic reasons, to control the tomato fruit worm?[4] Is it important to you to know that the tomatoes in your spaghetti sauce were beautiful before they were crushed? Is your health worth it?

Pesticides can help us control pests that carry or cause disease. We weren't very effective at controlling these pests before the advent of pesticides. Because of the use of fungicides (that reduce the growth of fungi), today's foods carry only traces of fungal toxins, among the most potent cancer-causers.

Economically, pesticides reduce crop losses and food prices. It has been estimated that, if we eliminated pesticide crop management, food losses and crop prices could be 50 percent greater.[3] Isn't it better to pay more now than to pay later—with your life?

Pesticides are used in hospitals, restaurants, warehouses, cinemas, and hotels (to name a few). They are found on grocery store, garden, farm, and feed store shelves. Crops are just some of the many products contaminated with pesticides. Others include house plants, wood products, and rubber products. But more critical—and dangerous to the public—is that pesticide residues are found in our food and in our water. The fact that fruits and vegetables are sprayed so many times during their growing period makes it difficult to believe the final product could be residue-free.

There are many ways we can be exposed to pesticides. These include eating foods containing pesticide residues, drinking contaminated water, exposure in various occupational and agricultural settings, and through pesticides used in the yard, home, and office.

Those at risk in agricultural settings include forestry, greenhouse, nursery, and lawn-care workers, as well as field workers and pickers, and pesticide applicators.

Approximately $7 billion worth of pesticides (about one billion pounds) are used annually in U.S. agriculture.[5] Worldwide use, though, tops four billion pounds![5] Many of these active pesticide ingredients have never been tested for potential neurotoxic or neurobehavioral effects, damage to the reproductive system, or other effects on human health. Few pesticides have ever been restricted or banned by the Environmental Protection Agency (EPA).

Do pesticides really do the work they're supposed to? Not necessarily. More and more pests are becoming resistant to pesticides. In fact, by 1984, over 400 species were found to be resistant to one (or more) insecticides.[2] The National Academy of Sciences has reported that some pests (such as the Colorado potato beetle in Long Island) are completely resistant to every chemical insecticide in existence.[2] Maybe it's time to find another solution.

How Pesticides Affect Enzymes

Pesticides, at the biochemical level, may affect humans in the same way they affect the insects for which they are intended: that is, through inhibition of the enzyme that breaks down the neurotransmitter *acetylcholine*.

The enzyme *cholinesterase* is critical to our nervous system, and is destroyed by pesticides. Messages from the brain to the heart are carried by a chemical known as acetylcholine. The parasympathetic nervous system can be overstimulated, leading to severe health problems, if too much acetylcholine accumulates in the gap between one nerve cell (neuron) and the next. This overstimulation affects us in many ways; it can stop the bowels from working, slow down the heart, and so on. Under normal conditions, a gap between these nerve cells exists. In a normal body, the enzyme cholinesterase clears the acetylcholine away (once the nerve has fired). Wholesale, slow poisoning of multiple enzyme systems occurs because our air, food, and water are polluted with pesticides.

Similar to carbamates and organophosphorous pesticides, all cholinesterase-inhibiting pesticides can cause hyperactivity, neuromuscular paralysis, visual problems, breathing difficulties, abdominal pain, vomiting, diarrhea, restlessness, weakness, dizziness, and possibly convulsions, coma, or death. This residue is in the soil, in the plants, and in you. Our government allows this.

The key to all life processes is enzymes. All chemical reactions in the body are regulated by the catalytic action of enzymes. Enzymes are highly specific in their actions and each catalyzes only a small range of chemical changes in its living cell. Any disturbance in ability to function will decrease the activity and strength of enzymes. This could result in illness and feelings of fatigue.

Enzyme systems are disturbed by pesticide toxins in a number of ways similar to nerve gases or other poisons. Eating only organic, pesticide-free food is one way to maintain and restore your enzyme systems. Remember, you're fighting for your life! The wisdom of spraying antienzymes (pesticides) on vegetables and fruits seems questionable, at the least.

Enzymes require vitamins and minerals in order to function. Food is the main source of these vitamins and minerals. It is extremely dangerous to spray growing crops with deadly chemicals, thus interfering with the plant's absorption of vitamins, minerals, and enzymes from the soil. In addition, processing and refining fruit, vegetables, and grains strips them of their vitamins and minerals. Therefore, people eating pesticide-sprayed, factory-produced food are consuming enzyme-deficient substances, void of nutrients, but containing possibly deadly poisons. Is this the way to jump start your day? Could the box containing your breakfast cereal be more nutritious than the cereal itself?

The Natural Resources Defense Council (NRDC) examined twenty-three pesticides known to have adverse health effects. They reported that preschoolers are being exposed to hazardous levels of pesticides in fruits and vegetables. Twenty of these pesticides were found to be neurotoxic. NRDC estimated that, from raw fruits and vegetables alone, at least 17 percent of the preschool population (or 3 million children) are exposed to neurotoxic organophosphorous pesticides above levels the federal government has described as safe.[6]

Unfortunately, there are major gaps in data in many pesticide registration files. In 1984, the National Academy of Sciences found that 67 percent of pesticides studied had undergone *no neurotoxicity testing at all.* In addition, all of the neurotoxicity tests performed were judged to be inadequate.[7] This should be frightening to you; it sure is to me.

More than 65,000 chemicals are in the EPA's inventory of toxic chemicals, and the agency annually receives approximately 1,500 notices of intent to manufacture new substances.[8] No precise figures are available on the total number of chemicals in existence that are potentially neurotoxic to humans, since few of these chemicals have been tested to determine if they adversely affect the nervous system.

What Can We Do?

What can we do to eliminate or control the use of pesticides? Immediately, four basic measures come to mind:

1. Make appropriate changes in food processing, preparation, and storage.
2. Farm organically.
3. Employ integrated pest management (i.e., using lady bugs as a natural enemy of aphids).
4. Use pesticides only when other measures are impossible.

Remember that processing and cooking procedures, as well as storage conditions can affect the amount of pesticide residue in food. Wash your fruits and vegetables thoroughly before you eat them. In one study, simply washing tomatoes reduced the residue of the pesticide benomyl by 80 percent.

HEAVY METALS AND OTHER TOXINS

Pesticides aren't the only things that can inhibit enzymatic activity. Lead and mercury are heavy metals, which in sufficient concentrations, can negatively impact your health. In part, this is because of what they do to your enzyme systems.

Lead

Exposure to lead can adversely affect your heart, nerves, blood, skin, metabolism, liver, kidneys, and reproductive organs, to name a few. Those who work around lead may expose their children and families to this toxic metal inadvertently by transporting lead dust into their homes on their clothing.

Pregnant women exposed to lead may give birth to premature babies with birth defects. Even maternal lead exposures below 25 mcg/dl can lead to lower child IQ, slower reaction time, inadequate vitamin D metabolism, reduced size up to eight years of age, and other neurotoxic effects.[9]

Highly elevated lead levels can cause coma, convulsions, profound irreversible mental retardation, seizures, and even death. Even low levels of exposure can result in persistent impairments to central nervous system function, especially in children, including delayed cognitive development, reduced IQ scores, impaired hearing, adverse impacts on blood production, vitamin D, and calcium metabolism (which have far-reaching physiological effects), and growth deficits. In adults, lead in the blood may interfere with hearing, increase blood pressure, and, at high levels, cause kidney damage and anemia.[10]

Lead (as an insoluble phosphate) deposits in the bones in much the same way as calcium. This effectively removes it from the bloodstream and detoxifies it. During strenuous exercise (or other periods of acidosis), this lead can be mobilized into the bloodstream and can cause lead poisoning.

Basic research on the health effects of lead now indicates that its toxicity is greater than previously thought and at considerably lower levels.

In 1979, more than 90 percent of food cans in the United States were sealed with lead solder. By 1989, that figure had decreased to 4 percent and by 1991, U.S. food canners had quit using lead solder. Unfortunately, the number of imported cans containing lead solder is not known but may be large.[11] A Food and Drug Administration (FDA) investigation of ethnic grocery stores conducted in California in 1997 found more than 100 cans sealed with lead solder.[12]

You should also be aware that lead could be in your pottery and in ceramic dishes, especially if they were foreign-made. Most dinnerware is coated with a lead glaze. If improperly applied and fired, however, lead may leach out of the container and into your food. Many countries lack the stringent regulations we have regarding lead in ceramics, so be careful with imported dishes.

High blood lead levels are a major environmental threat to the health of American children. Childhood lead poisoning is completely preventable. Decreased levels of lead in gasoline, air, food, and industrial pollution have helped. However, lead can still be found in paint, dust, and soil (especially in inner-city urban areas).

The primary source of lead in drinking water is corrosion of plumbing materials, such as lead service lines and lead solders, in water distribution systems, in houses, and larger buildings. Virtually all public water systems serve households with lead solders of varying ages. What's worse, most faucets are made of materials that can contribute some lead to drinking water. Unfortunately, our homes are still the major source of lead exposure in the United States.

Enzymes involved in the synthesis of iron (necessary for the formation of red blood cells) are poisoned by lead. The body compensates for this by manufacturing more red cells, which tend to be imperfect. Excessive lead in the environment causes significant chemical alterations in humans, and therefore, should be avoided.

Mercury Menace

Similar to lead, mercury compounds are potent neurotoxic substances and have caused a number of poisonings worldwide. Common symptoms of exposure include lack of coordination, speech impairment, vision problems, tremors, headaches, and nausea. Severe cases of mercury poisoning have been linked to brain damage, kidney disease, and death.[13]

Mercury occurs naturally in almost all foods, but within normal limits. The exception to this is seafood, which because of polluted oceans, can show concentrations greater than normal.

In the mid-1950s, a chemical plant near Minamata Bay, Japan, discharged methylmercury (a highly toxic organic form of mercury) into the bay as part of waste sludge. Fish and shellfish became contaminated. Local residents, after eating the seafood, experienced severe mercury poisoning and its accompanying neurotoxic and developmental effects.

In another incident, mercury (used as a fungicide in treating seed grain) was the cause of a very serious epidemic in Iraq in 1971. More than 450 people died from eating the treated grain.[14]

Drugs as Pollutants

Not only is there pollution in our food, air, and water, but also in the very drugs that are prescribed for us to fight deadly diseases, such as cancer. Is it possible the very people we go to for lifesaving assistance are inadvertently giving us drugs that cause cancer and other killer diseases?

A list of over thirty-seven anticancer drugs with reported carcinogenicity or cocarcinogenicity was reported by Eric J. Lien and Xingchang Ou (School of Pharmacy, University of Southern California). These drugs include sixteen alkylating agents, eight antimetabolites, four antibiotics, four hormones, one alkaloid, and four miscellaneous drugs. These drugs (which include azathioprine, progestin, and testosterone propionate) have been reported to cause cancers in different sites in various test animals. In addition, most of these drugs react with DNA.[15]

Summary on Toxins

It seems critical to change our lifestyles by avoiding pesticides, heavy metals, and "killer" drugs. We should also reinforce our bodies by using natural health therapy to eliminate toxins and to build up our poisoned systems.

As you can see, the pesticides, heavy metals, and killer drugs in our environment affect our enzyme systems. They interfere with the minerals, vitamins, and enzymes that our foods absorb from the soil. We, in turn, suffer from that deficiency. To make matters worse, our food processing methods can also affect the enzyme levels in our foods.

In order to make our systems run better, we need to eliminate the sludge, the garbage present in our bodies. To get the garbage out we have to detoxify. This is the subject of our next chapter.

References

1. *Neurotoxicity*. Washington, D.C.: Office of Technology Assessment, 1990, 49.
2. Michael F. Jacobson, Lisa Y. Lefferts, and Anne Witte Garland, *Safe Food*. New York: Berkley Books, 1993, 45–47.
3. Val Hillers. Washington State University College of Agriculture and Home Economics, 1990.
4. *Safe Food*, 59.
5. *Neurotoxicity*, 282.
6. February, 1989 report, "Intolerable Risk: Pesticides in Our Children's Food," published by the Natural Resources Defense Council (NRDC).
7. *Neurotoxicity*, 297.
8. *Neurotoxicity*, 3.
9. *Healthy People 2000*. Washington, D.C.: U.S. Department of Health and Human Services, Public Health Service, 1991, 303.
10. Ibid. 320.
11. *Neurotoxicity*, 275.
12. *FDA Consumer*, Washington, D.C.: U.S. Food and Drug Administration. January/February 1998. Retrieved from http://vm.cfsan.fda.gov/~dms/.
13. *Neurotoxicity*, 131.
14. Ibid. 48.
15. Eric J. Lien and Xing-chang Ou, "Carcinogenicity of Some Anticancer Drugs—A Survey," *Journal of Clinical and Hospital Pharmacy* 10:223–242 (1985).

CHAPTER 5

Garbage in—Garbage out

FEEL SLUGGISH? Run down? Too pooped to pop? Dragging? Is that tube (the alimentary canal) that runs the entire length of your body plugged somewhere along the trail? If so, there is probably a sludge buildup. The answer, my friends, is *detoxification*.

DETOXIFICATION

The role of detoxification is the removal of toxins from the body, no matter what their origin. Sometimes these toxins are actually the *reason* for a disease process.

As noted earlier, toxins are substances harmful to our bodies. They can develop from our external environment or be produced from within us. External toxins known to cause significant health problems include heavy metals, drugs, microbial toxins, pesticides, and solvents. Concerning heavy metals, it is conservatively estimated that up to 25 percent of the U.S. population suffers in varying degrees from heavy metal poisoning.[1] These metals include aluminum, arsenic, cadmium, lead, mercury, and nickel. They can severely disrupt normal function and tend to accumulate in the kidneys, immune system, and brain.

When toxins originate from within, it is because of the body's inability to rid itself of metabolic waste products, incomplete oxidation in the tissues, inadequate nutrition, and/or because of some existing disease process. Our bodies have a big job getting rid of toxic substances, regardless of their cause. There is a heavy demand on the liver, lungs, kidneys, and skin to help eliminate any toxic substance.

Detoxification is basically a twofold process. The first is to avoid environmental toxins as much as possible and the second is to aid the body in voiding itself of toxins (whether due to environmental factors or internal disorders). The human body possesses many methods of detoxification and regeneration. Nature is resourceful, simultaneously using several systems for the maintenance or re-establishment of health. Enzymes play an important role in this process.

When our body's self-healing powers are overtaxed by toxins (both external and internal), this leads to stress on the immune system. This, in turn, can lead to chronic degenerative disorders. Natural treatment methods of detoxification, self-cleaning, and elimination can successfully overcome our inability to cope with infection and degenerative disorders, thereby invigorating our immune system. The use of certain enzyme mixtures has proven to be especially effective in detoxification, elimination, and activation of the body's immune defenses. This process is capable of fighting even chronic disorders such as cancer and rheumatism.

A successful detoxification program should involve: changes in eating habits, fasting, fresh juices, therapeutic exercise, hydrotherapy, and, of course, enzyme treatment.

What Happens During Toxin Buildup?

Foreign substances and pollutants (including alcohol, tobacco, and drugs) continually subject our bodies to increased stress. Our modern lifestyle, with unhealthy eating habits (especially excessive intake of refined carbohydrates and fat), lack of exercise, and daily stress builds up toxins in our bodies. All too frequently, many of us pop a handful of tranquilizers or pain killers into our mouths. We overreact to any condition, headaches, infections, digestive complaints, and the like. We want to be healed yesterday. We erroneously treat viruses with antibiotics and anti-inflammatory drugs, thus paralyzing our defense mechanisms, suppressing the natural course of self-healing. This encourages a disease to become chronic. Since the body has many methods of detoxification and ways to regenerate itself, our job is to help it help itself.

We need to be detoxified frequently in order to better absorb nutrients from food and eliminate toxins. This is because of the tremendous amount of chemical food additives, such as preservatives, artificial colors and flavors, emulsifiers, stabilizers, and sweeteners in our foods and beverages. During digestion, nutrients are metabolized forming water, energy, waste products, and carbon dioxide. Frequently, the accumulation of these wastes interferes with normal cell function—hence the need to detoxify.

The Body's Self-Healing Powers

We continually excrete large quantities of wastes, toxins, and dangerous metabolic products through our lungs, intestines, kidneys, and skin. The body strives to maintain good blood and lymphatic circulation at all times. Blood from the arteries not only brings oxygen and nutrients to maintain cell metabolism, but it also cleans out the cells. It does the garbage disposal work by transporting and eliminating metabolic debris and waste products.

A portion of this disposal occurs in the breakdown (catabolism) and detoxification of waste in organs such as the lymph, liver, and spleen. These broken-down waste products are then excreted by way of the kidneys, intestines, lungs, and skin.

Therefore, the primary goal of detoxification and natural health care is to assist in the breakdown, elimination, and excretion of toxins and wastes from the body; to give relief to a depressed immune system; and to stimulate or invigorate our detoxification mechanisms.

How do we maintain the body's excretion mechanisms? One of the main keys to a healthy body is a properly functioning intestine. Of utmost importance is keeping the colon clean. Constipation, diarrhea, and similar conditions can be related to many diseases, as well as defective colon function. It is critical for a healthy body that we keep the colon clean and free of debris. The stress on the body is incalculable when toxins are reabsorbed (from food putrefaction) into the overburdened bloodstream and taken to the already overworked liver. It is always surprising to see how much better many chronically ill patients become when normal intestinal activity is restored. This applies not only to conditions of digestion and absorption, but also to chronic disorders of other systems and organs, not directly related to bowel function (such as circulation, kidney, and liver problems). In addition, digestive problems can accelerate aging, and trigger allergies, skin blemishes, and so on.

The Liver and Gallbladder Flush

An important detoxifying tool is the liver and gallbladder "flush." This flush can help restore the normal functional capacity of the liver and gallbladder, especially in those suffering from chronic degenerative diseases. This flush is recommended for adults *only*. Check with your physician before attempting.

Individuals should follow these directions:

1. On days one through six, in addition to supplements and regular meals, drink as much apple cider (or juice) as the appetite will permit. To assure there are no additives, the juice or cider should be from organically grown apples.
2. Eat a normal lunch at noon on day seven.
3. Take 2 teaspoons of disodium phosphate (dissolved in about one ounce of hot water) about three hours later. Because the drink might be unpleasant, it may be followed by a little citrus juice (if possible, freshly squeezed). Those on sodium-free diets may substitute a sodium-free cathartic in place of the disodium phosphate (i.e., Epsom salts, castoria, etc.). Ask your health-food store or doctor for sources of disodium phosphate.
4. Repeat step three two hours later.
5. Take grapefruit juice, grapefruit, or other citrus fruits or juices for the evening meal.
6. At bedtime, take either (a) ½ cup warm, unrefined olive oil, blended with ½ cup lemon juice or (b) ½ cup of unrefined olive oil followed by a small glass of grapefruit juice. Any health-food store should sell unrefined olive oil. Though canned or bottled citrus juices are permissible, fresh is best.
7. Immediately after this, go to bed and lie on the right side with your right knee pulled up close to your chest (for thirty minutes).

8. One hour before breakfast (the next day), again take 2 teaspoons of disodium phosphate dissolved in 2 ounces of hot water.
9. After this, return to your normal diet and any nutritional program that may have been prescribed.

Slight to moderate nausea may occur when taking olive oil/citrus juice. This nausea should slowly disappear by the time you go to sleep. If the olive oil induces vomiting, you need not repeat the procedure at this time. Nausea only occurs in rare instances. The liver and gallbladder are stimulated by this flushing and these organs are cleaned.

The following day, chronic sufferers from gallstones, backaches, and nausea may find small gallstone-like objects in the stool. There may be a large number of irregularly shaped, gelatinous objects, varying in size (from grape seed to cherry seed size). Patients who find these gallstone-like objects should repeat the flush in two weeks.

The Kidney Flush

Approximately 4,000 quarts of blood are filtered daily by the kidney. The elimination of waste (mostly urea) helps maintain the acid/alkaline balance of the kidney. Ample amounts of liquids, such as distilled or spring water, should be taken daily. If you are nutritionally deficient, however, vegetable and fruit juices should be used in addition to raw, whole fruits and supplements. Protein intake in kidney disease should be limited because of the stress protein places on the kidney.

By using the following method, the kidney can be detoxified: Drink as much watermelon juice or eat as much watermelon as possible. Juice can be made in a juicer or blender. Because watermelon's availability is usually limited to the summer months, you can substitute the following methods:

1. Drink six ounces of water mixed with the juice from one lemon. If needed, use ½ teaspoon of honey to improve the flavor.

2. For one week, drink two glasses of unsweetened cranberry juice per day.

Purge

Take 1 tablespoon of Epsom salts (or 1 teaspoon of disodium phosphate) dissolved in a half-glass of distilled water immediately upon arising. Repeat this same procedure two more times, waiting thirty minutes between each time. This procedure should be repeated three times over ninety minutes (in other words, every thirty minutes).

If you become hungry, the following alkalizing punch may be taken after one to two hours, or as desired.

Alkalizing Punch

Place the juice of twelve oranges, six grapefruits, and six lemons in a gallon glass jug filled with distilled water. Drink a glass hourly, or as desired.

The purge should be repeated for one additional day or a total of two days every four weeks, if possible. If the purge weakens you too much, discontinue, or repeat only every three to four weeks. The body may undergo some rather uncomfortable sensations during the purge, such as nausea, headaches, cramps, and dizziness. Toxins have accumulated for years and are suddenly being dumped; this can result in a shock to your system. These sensations (which are temporary) should disappear when the toxins are finally eliminated. Every three or four weeks, repeat the purge.

Again, this procedure should be used on the advice of an experienced and well-trained health-care professional.

Retention Enema

A retention enema can be used as part of a detoxification program. Water, coffee, or other substances are used in these enemas, which are retained approximately fifteen minutes before being voided. Check with your physician.

Key Points in Detoxification

A great deal can be achieved by restorating intestinal flora and reestablishing intestinal peristalsis. Various methods of cleansing include fasting, juicing, detoxification diets (including high-fiber, enzyme-rich diets), exercise, and hydrotherapy. See chapter 8 for details. Colonic irrigation should be used only in exceptional cases, and then only under the supervision of a well-trained physician. Make haste slowly when considering colonic irrigation.

These techniques can help chronic kidney and rheumatic disorders, depressions, unbearable headaches, as well as other conditions. Concurrently, there should be biological stimulation of kidney function. This goal can usually be achieved by increased liquid intake and mild herbal diuretics. Check with a nutritionally oriented health-care provider.

It is important to review our food and drinking habits, plus where we eat, when we eat, and with whom. Toxins such as caffeine, drugs, and alcohol should be avoided, as well as all those in our general environment and workplace.

"An ounce of prevention is worth a pound of cure." It is very important to correct any problems which can cause or contribute to toxic disorders (i.e., infected gums, teeth, tonsils, etc.).

Supplements should also be used regularly to rebuild our defense barriers, such as mucosa and skin. Antioxidants, including vitamins A, E, and C, selenium, zinc, the enzyme superoxide dismutase (SOD), and digestive and other enzymes are all essential. In addition, certain enzyme mixtures also provide support for circulation to (and function of) the mucosa; promote waste detoxification and disposal; help to heal chronic infections; and accelerate the inflammatory process, thus expediting return to healthy tissue.

The primary goal of detoxification is to:

1. Break up wastes and toxins in the body
2. Eliminate wastes and toxins
3. Improve excretion
4. Invigorate the body's detoxification mechanism
5. Provide relief when the immune system is depressed

Reference

1. Michael Murray and Joseph Pizzorno, *Encyclopedia of Natural Medicine.* Rocklin, Calif.: Prima Publishing, 1991, 32–40.

PART TWO

Energizing Enzyme Tools

CHAPTER 6

Enzymes' Coenergizers: Vitamins and Minerals

AS WITH any complex machine, the human body works best when everything is doing its job. But nothing in the body works alone. Most systems require "helpers" to do their work properly. Vitamins and minerals are enzyme helpers. Without them, most enzymes could not function. Because of their importance to enzyme activity, they are classified as *coenzymes* and *cofactors*.

COENZYMES

What are coenzymes? A coenzyme is necessary for an enzyme to function. Our bodies cannot make coenzymes; they *must* be obtained from our food. Coenzymes, in the human, are usually the B vitamins. The B vitamins thiamin, riboflavin, niacin, pantothenic acid, and biotin serve as helpers to enzymes which release energy from carbohydrates, fat, and protein. Vitamin B_6 assists those enzymes that metabolize amino acids, while folate and B_{12} work with enzymes in helping cells to multiply.

Fat-soluble vitamins are not usually considered to be coenzymes. But because they are involved in so many bodily functions, their presence is essential. For this reason, let's consider them as *indirect coenzymes*.

Unlike enzymes, coenzymes are not proteins. In addition, because coenzymes are consumed when performing their functions, *they require constant replacement.*

Sometimes, an enzyme requires a mineral or an electrolyte to facilitate a chemical reaction. These are called *cofactors.* An enzyme that contains one or more minerals as part of its structure is known as a *metalloenzyme.*

Vitamins

Vitamins are organic (that is, carbon-containing) compounds that are essential in very small amounts for health, growth, and reproduction. Most vitamins must be obtained from our diet or as supplements because either the body can't make them at all, or it can't make them in sufficient amounts. Vitamins are classified according to their solubility in fat or water, which affects their occurrence in foods, as well as their absorption, transport, storage, and metabolism in the body.

Fat-Soluble Vitamins

The fat-soluble vitamins are vitamins A, D, E, and K. These vitamins are usually found in high concentrations in the fatty portions of food. They are also absorbed, transported, metabolized, and stored along with fat. To be properly absorbed, they require bile and dietary fat (so a deficiency in either will cause absorption problems of these vitamins). They are transported in the body through the same mechanisms by which fat is transported and are stored in liver and fat tissue. Fat-soluble vitamins are excreted into the intestine in bile and are either reabsorbed or are eliminated in feces. They are not excreted (in any amount)) in urine. Because excretion of these vitamins is minimal, excess intake can cause toxicity symptoms. Deficiencies can occur among children who are growing rapidly and who lack adequate fat stores, and among children or adults who have disease conditions that interfere with fat metabolism, such as malabsorption, biliary obstruction, or renal or liver disease.

Vitamin A is present in the diet both as the vitamin, and as its precursor, retinol, or preformed vitamin A. It is found in foods derived from animals (such as milk, butter, egg yolks, liver) and, when bound to a fatty acid, is used to fortify many foods. Certain carotenoids (pigments found in many dark green, yellow, and orange vegetables and fruits, and egg yolks) can be converted by the body into retinol. The conversion of beta-carotene into retinol occurs mainly in the intestinal mucosa. Retinol circulates in the plasma bound to a specific transport protein called *retinol-binding protein.*

Excess amounts of vitamin A are stored in the liver and excessive intake has caused toxic symptoms, such as headache, skin and bone disorders, and renal failure. High intakes of retinol supplements have also been associated with birth defects, while synthetic retinoid analogs (used to treat a variety of skin disorders) can cause fetal malformations. They are hazardous to pregnant women or women planning to become pregnant and should be used only under medical supervision. Excess amounts of beta-carotene are stored in body fat deposits, though intake of foods rich in beta-carotene, such as carrots, is not known to cause toxic effects. Because it raises levels of carotene in the blood, it can cause the skin to take on an orange color. The color will disappear when the carotene consumption declines.

Vitamin A is essential for visual processes, for the normal differentiation of epithelial tissue, for the regulation of cell membrane structure and function, and for the maintenance of immunocompetence. Vitamin A deficiency (through adverse effects on eye epithelial tissue) is a major cause of blindness among children in many developing countries, and it is also responsible for substantial additional illness.

To be absorbed, vitamin A requires a certain level of fat in the diet and adequate quantities of *pancreatic lipid-digesting enzymes* (lipases) and bile salts in the small intestine. So, if you don't have adequate levels of dietary fat or enough lipases and bile salts, you may not absorb the vitamin A in your diet. In addition, some complications of alcoholism, such as pancreatic insufficiency and biliary insufficiency, can lead to vitamin A malabsorption. Alcoholics with liver disease may have impaired storage or transport of vitamin A because of an inadequate synthesis of retinol-binding protein (the protein formed in the liver that transports vitamin A in the blood). Even moderate alcoholic liver disease is associated with severely decreased vitamin A concentrations; these levels are reduced in the liver even when blood levels of vitamin A, retinol-binding protein, and prealbumin are normal.

The storage form of vitamin A, retinol, is oxidized to its active form, retinal, by *retinol dehydrogenase*, an enzyme similar to alcohol dehydrogenase. As mentioned above, impairments in the metabolism of vitamin A have been reported in alcoholics, and some evidence suggests that retinol and ethanol compete for the retinol dehydrogenase enzyme in the liver, testes, and retina.[1] If your body's enzyme, retinol dehydrogenase, is being used to metabolize the alcohol you ingest, then how can it do its work to help your vitamin A stores?

Scientists are also studying vitamin A and beta-carotene-containing foods in cancer epidemiology. Experimental deficiencies of several nutrients, including vitamin A, have been shown to enhance carcinogen-induced tumors in laboratory animals. The role of vitamin A deficiency in the reported association between alcohol intake and certain types of tumors still remains to be determined.

A large body of evidence suggests that foods high in vitamin A and carotenoids can protect us against a variety of epithelial cancers. Vitamin A in foods occurs in two forms: (1) preformed vitamin A (retinol and retinol esters) from animal foods and (2) provitamin A (carotenoids that the body can convert to vitamin A) from plant foods. Although beta-carotene is the most efficiently converted, it still has only one-sixth the biologic activity of retinol.

Low levels of vitamin A (or retinol) in blood have been associated with an increased risk for cancer in some studies, as have low levels of beta-carotene. There are problems, however, in relating blood levels to cancer risk. Blood retinol remains unchanged across a wide range of dietary intakes because of homeostatic mechanisms. Thus, increased vitamin A is unlikely to result in increased serum retinol, except during vitamin A depletion. Consequently, the association between serum retinol and cancer may be due to factors that regulate serum retinol rather than to the dietary vitamin itself. In contrast, serum carotenoids may better reflect dietary carotenoids, and the relationship between dietary carotenoids (or the foods that contain them) and cancer may be more direct than that between serum retinol and cancer. A major weakness of these studies is that blood levels of vitamin A or carotenoids observed when cancer is finally diagnosed may be different from those present when cancer actually began, so it's difficult to draw accurate conclusions.

Vitamin D_3 (cholecalciferol or calciol) is synthesized from a precursor (7-dehydrocholesterol) in skin that activates when we are exposed to the sun's ultraviolet light. It is essential in our diet at times when our exposure to the sun is inadequate. The vitamin is converted by the liver to calcidiol, and then further converted by the kidney to calcitrol, the metabolically active form. Excess vitamin D_3 can be toxic, especially to children and adults who have kidney disease or certain metabolic disorders.

Vitamin D_3 and its metabolites are involved in regulating calcium and phosphorus metabolism, bone formation and resorption, and various other physiologic functions, including some aspects of immune function. Alcoholics have been reported to have low circulating levels of vitamin D_3, especially calcidiol, the metabolite formed in the liver. Not surprisingly, alcoholics with liver disease have lower levels of this metabolite than persons without liver disease. Higher rates of osteomalacia and osteoporosis have also been reported in alcoholics, possibly because a vitamin D_3 deficiency would lead to problems in calcium absorption. In addition, some evidence suggests that alcohol induces *enzymatic changes* in the liver that favor the production of inactive metabolites of vitamin D_3. The actions of vitamin D_3 are closely linked with the metabolism of calcium and phosphorus. In fact, their levels may be altered in response to heavy alcohol intake.

Vitamin E functions as a lipid-soluble antioxidant and free-radical scavenger. Thus, the protective role tentatively assigned to both carotenoids and vitamin C may also apply to vitamin E and its derivatives. Vitamin E, like vitamin C, blocks the in vitro formation of nitrosamines. The fact that vitamin E is lipid (fat) soluble permits this action in a lipid environment, as opposed to vitamin C, which is water soluble.

Its principal dietary sources are vegetable seed oils. A deficiency of this vitamin has been associated with hemolytic anemia in premature infants and with neurologic symptoms in adults.

Vitamin K functions as an activator of blood-clotting proteins, proteins in bone and kidney, and the formation of other proteins that contain gamma-carboxyglutamic acid (GLA). Its main function is the coagulation of blood. Because vitamin K is synthesized by intestinal bacteria, deficiencies generally occur only in infants whose intestinal flora has not yet been established, in children and adults receiving antibiotic or anti-coagulant therapy, and in individuals with disease conditions that interfere with intestinal absorption. Vitamins E and K are less toxic than vitamins A or D.

Some substances can "mimic" vitamin K and are mistaken for it by the body, leading to disastrous consequences. When this happens, vitamin K can't do its work (as a coenzyme), leading to illness and possibly death.

An application of this concept is the rat poison, warfarin. Because the body mistakenly sees this poison as vitamin K, it will incorporate it into the enzymes (instead of vitamin K). Because vitamin K is necessary for blood coagulation, a deficiency in this vitamin can lead to uncontrolled internal bleeding. This is exactly what happens when rats (or mice) ingest warfarin; their bodies fail to use vitamin K in the enzymes. This eventually leads to internal bleeding and death. Small amounts of warfarin, such as in the drug Coumadin®, however, can be used in humans as an anticoagulant (an effect desired in some cardiovascular patients).

Evidence suggests that at least half of the vitamin K required by humans is normally synthesized by bacteria in the intestine. In rare cases, vitamin K deficiency may occur with fat malabsorption, which is common in alcoholics.

Water-Soluble Vitamins

The water-soluble vitamins include vitamin C (ascorbic acid), and those of the B-complex group: biotin, folate, vitamin B_3, pantothenic acid, vitamin B_2, vitamin B_1, vitamin B_6, and vitamin B_{12}. These vitamins are generally found in whole grain cereals, legumes, leafy vegetables, and meat and dairy foods. The two exceptions are vitamin C, which can be obtained in adequate amounts only from fruits and vegetables, and vitamin B_{12}, which is made by bacteria and found only in foods of animal origin.

Water-soluble vitamins are absorbed from the intestine, and most are stored in a form that is bound to *enzymes* or transport proteins and excreted in urine. Thus, they should be supplied in adequate amounts in the daily diet.

Water-soluble vitamins are essential components of enzymes and enzyme systems that catalyze a wide variety of biochemical reactions in cellular energy production and biosynthesis. Deficiencies of these vitamins can keep your enzyme systems from working properly. Deficiencies can particularly affect tissues that grow or metabolize rapidly, such as skin, blood, and the cells of the digestive tract and nervous system. Common deficiency symptoms are skin disorders, anemia, malabsorption and diarrhea, neurologic disorders, and defects in tissues of the mouth. When deficiencies occur, they are usually found along with other vitamin deficiencies and are due to diseases, to consumption of highly restricted diets, consumption of excessive amounts of alcohol, or intake of drugs that interfere with vitamin metabolism (and many drugs can interfere). The risk for deficiencies is greater in growing infants and, perhaps, in older persons.

Vitamin C functions as a chemical-reducing agent and antioxidant. It is synthesized in adequate amounts by most animals, *but not by man* (or some primates, fish, or guinea pigs). So *all* of our vitamin C must come from our diet or from supplements.

Human studies have shown a protective association between foods that contain vitamin C and cancers of the esophagus, stomach, and cervix. Small-scale studies have indicated that colonic polyps regress or decrease in area with vitamin C therapy. Recurrence of colonic polyps after their surgical removal was reduced among patients in the treatment group of a study with 200 subjects.[2] While many studies support a role of vitamin C in reducing cancer risk, no wholly consistent picture of the role of vitamin C in human cancer has been defined.

Biochemical studies suggest that vitamin C blocks the formation of carcinogenic nitrosamines from nitrates and nitrites within the digestive tract and prevents oxidation of certain chemicals to active carcinogenic forms.

Just as it affects every other vitamin, alcohol intake can affect your vitamin C levels, too. Serum ascorbic acid levels have been found to be lower in alcoholics than in nonalcoholics, but this may be because alcoholics don't eat properly. However, some data suggest that even when diets are adequate, increasing levels of alcohol consumption are associated with lower serum levels of ascorbic acid.

Vitamin C is essential for the proper functioning of the enzyme that hydroxylates proline (an amino acid used to make collagen). Collagen is the protein material from which connective tissues (such as ligaments, tendons, scars, and the foundation of bones and teeth) are made.

The list of **B vitamins** is extensive and includes thiamine (B_1), riboflavin (B_2), biotin, folate, niacin (B_3), pantothenic acid, vitamin B_6, and vitamin B_{12}. Most B vitamins serve as coenzymes to various enzyme systems. Their presence is critical for the proper functioning of the enzyme systems. Even if you have an adequate enzyme supply, your enzymes might not be doing what they're meant to do if you don't take in enough B-complex vitamins.

Without B vitamins you will lack energy. This is not because of the vitamins themselves, but because some of the B vitamins act as enzyme helpers to release energy from the carbohydrates, fats, and proteins that you eat. Other B vitamins help cells to multiply. This is especially important in cells that have short life spans, such as red blood cells and the cells that line the GI tract. These cells help deliver energy to the body's tissues and must replace themselves rapidly.

Every single one of the B vitamins plays a role in metabolism. Energy metabolism is a catch-all phrase that describes all the ways your body obtains and uses energy from the food you eat. As you know, you eat mostly carbohydrates, fats, and proteins which are broken down by the body for use. The breakdown is done in a series of stages, each one dependent on the previous stage. These are called "metabolic pathways." To better understand this concept, let's assume you can only enter your house by opening a gate, walking up the sidewalk, and entering through the front door. If you can't get the gate open, you'll never make it up the walk and into the front door. If you can't get the front door open, you'll never get into the house. Metabolism works in much this same way. Each stage of the pathway must be completed before the next stage can be attempted. Enzymes make the completion of each step in the pathway possible.

For instance, thiamine's coenzyme form, *thiamine pyrophosphate*, is involved in energy metabolism and assists many enzymes, including pyruvate decarboxylase, alpha-keto acid decarboxylases, transketolases, and aldehyde transferases. If you don't get enough thiamine in your food, the above enzymes can't do their jobs. To make it worse, you may be doing things that interfere with absorption of the thiamine you do eat. For instance, antacids can impair thiamine absorption, as can certain drugs, any disease that affects intestinal absorption, and lack of sodium (however, most Americans eat too much, rather than too little, sodium).

Table 6–1 illustrates some of the B vitamins and their coenzymes.[3]

The B vitamins are all interdependent. Each one affects the absorption, metabolism, and excretion of the other B vitamins. For example, because folate is involved in thiamine absorption, a deficiency in folate will cause a thiamin deficiency. In turn, a deficiency in B_{12} (needed to free folate) will keep folate from manufacturing red blood cells.

Poor dietary intake and poor selection of foods will reduce thiamine intake. Alcohol may interfere with absorption of thiamin, with its activation to thiamine pyrophosphate, or with the ability of thiamine pyrophosphate to combine with the enzymes for which it is a cofactor.

Though alcohol affects all B-complex vitamins, folate deficiency is probably the most common vitamin deficiency observed in alcoholics. Conversely, alcoholism is probably the most common cause of folate deficiency in the U.S. adult population.

Urinary excretion of folate is increased by alcohol intake, and its tissue utilization is decreased. Alcohol may directly inhibit *enzymes* involved in folate metabolism. Several investigators[4,5] have shown that alcohol antagonizes the ability of folate to reverse the megaloblastic bone marrow changes seen in deficiency states, but larger doses of folate can overcome this antagonism. The suppressive effects of alcohol have been shown to be present whether folate is given intravenously or orally, suggesting that a metabolic function subsequent to absorption is involved.

In addition to alcohol, several drugs used as cancer chemotherapeutic agents (e.g., methotrexate), as diuretics (triamterene), or as antimalarial or antibacterial drugs (pyrimethamine) are antagonists of folic acid. These drugs bind to the enzyme *dihydrofolate reductase*,

Table 6–1. B Vitamins and Their Coenzymes

VITAMIN	COENZYME	GOOD FOOD SOURCES
Thiamine (B_1)	Thiamin pyrophosphate (TPP), a coenzyme in energy metabolism.	Whole grains, pork, nuts, eggs, legumes.
Riboflavin (B_2)	Flavin mononucleotide (FMN) and flavin dinucleotide (FAD), both coenzymes in energy metabolism.	Milk, cheese, eggs, whole grains, meats, leafy green vegetables, legumes.
Vitamin B_6	Part of the coenzymes pyridoxal phosphate (PLP) and pyridoxamine phosphate (PMP), used in fatty acid and amino acid metabolism.	Liver, fish, nuts, wheat germ, yeast, meats, whole grains.
Niacin (B_3)	Nicotinamide edenine dinucleotide (NAD) and the phosphate form (NADP), both enzymes in energy metabolism.	Liver, yeast, poultry, meats, fish, legumes.
Folate	Part of the coenzymes dihydrofolate and tetrahydrofolate (DHF and THF) used in DNA synthesis.	Whole grains, meat, poultry, fish, fruits, vegetables.
Vitamin B_{12}	Part of the coenzymes methylcobalamin and deoxyadenosylcobalamin, used in new cell synthesis. It also helps reform folate coenzyme.	Meat, fish, milk, eggs, shellfish.
Biotin	Part of a coenzyme used in energy metabolism, amino acid metabolism, fat synthesis, and glycogen synthesis.	Meat, milk, poultry, legumes, whole grains.

VITAMIN	COENZYME	GOOD FOOD SOURCES
Pantothenic Acid	Coenzyme A and others used in energy metabolism.	Legumes, meats, poultry, dairy products, whole grains, vegetables.

preventing the conversion of folic acid to tetrahydrofolate, the vitamin form that is required for synthesis of purines.

Also known as lipoic acid or ubiquinone, **Coenzyme Q10** is one of ten types of CoQ. Obtained in the diet (in organ meats, whole grains, nuts, seeds, and oily fish), CoQ10 is also made in our bodies and is naturally occurring in animals and plants. Plants also have coenzymes 6 through 9.

Dr. Emile Bliznakov, scientific director of the Lupus Research Institute in Ridgefield, Connecticut, has extensively studied Coenzyme Q10. In one study, he was able to extend the maximum lifespan of mice by using CoQ10.[6] Dr. Bliznakov took two groups of old mice. The control group (not fed CoQ10) died within nine months of the beginning of the study from natural causes. The experimental group (given CoQ10) were not all dead until more than twenty months after the experiment started. Not only did these mice live more than twice as long as the control group, they lived longer than mice normally do, and they had brighter eyes, more shiny hair, and lacked the normal signs of advanced age!

Coenzyme Q10 is present in live foods. But if you eat a modern, overly processed diet, you may be deficient in CoQ10. Coenzyme Q10 can be synthesized in the body from the amino acids tyrosine and phenylalanine, vitamin E and three B vitamins (B_1, B_6, and folic acid). Beef heart, organ meats, egg yolk, liver and whole grains are rich in CoQ10.[7]

Studies have been conducted around the world on CoQ10 and indicate that it may be helpful in generating energy, preventing heart disease and strengthening the immune system.

Minerals

Minerals perform a number of roles in the body. Certain minerals function as cofactors, inorganic components of *enzyme systems* that catalyze the metabolism of protein, carbohydrate, and lipids. Some act to regulate fluid and electrolyte balance and provide rigidity to the skeleton, while others regulate the function of muscles and nerves. Minerals also work together with vitamins, hormones, peptides, and other substances to regulate the body's metabolism.

Essential minerals are often classified as microminerals or trace elements. Microminerals are required in amounts from several hundred milligrams to 1 or more grams a day (calcium, phosphorus, magnesium, sodium, potassium, and chloride). Trace elements, such as iron, zinc, iodine, copper, manganese, fluoride, chromium, selenium, molybdenum, and cobalt (as a component of vitamin B_{12}), are required in small amounts. Other minerals, such as nickel, vanadium, silicon, or boron have been shown to be essential under rigorous conditions for experimental animals but do not have well-established functions in humans. Still others, such as lead or mercury, are potentially toxic.

Minerals are distributed in a variety of foods, but they usually are present in limited amounts. Thus, diets must contain a variety of foods to meet daily requirements. If you consume a low-calorie diet for a prolonged period, you will be at risk of developing mineral deficiencies. In addition, medications can interfere with mineral absorption and metabolism, leading to deficiencies. Alcoholism, renal disease, or gastrointestinal diseases can also cause deficiencies. Toxic symptoms can result from consumption of excessive amounts of almost any mineral or as a result of defective regulation or absorption or inadequate excretion.

Zinc is an essential trace mineral found in every tissue and tissue fluid of the body. Of the trace minerals, only iron is found in greater concentrations. Zinc is required as a cofactor involved in more than 300 enzymatic activities. These zinc metalloenzymes account for virtually every aspect of the metabolism of animal and plant foods.

Zinc has numerous functions in the body, as explained below:

1. Assists in white blood cell immune function
2. Assists in growth and development
3. Works with insulin in the pancreas
4. Interacts with platelets in blood clotting
5. Affects thyroid hormone function
6. Affects behavior and learning
7. Is needed to produce retinol (the active form of vitamin A) in visual pigments
8. Essential to normal salt-taste perception
9. Essential to wound healing
10. Essential in the production of sperm
11. Essential in fetal development
12. Helps protect the body from heavy metal poisoning

Zinc assists the various enzymes that metabolize carbohydrates, alcohol, and essential fatty acids. It works with the enzymes that synthesize proteins, dispose of free radicals, manufacture heme (a constituent of hemoglobin), and make components of DNA and RNA. Zinc also works with the pancreatic enzymes. These enzymes have some pretty important jobs and they can't do them without the assistance of zinc.

Reduced concentrations of zinc have been found in the plasma, red blood cells, and liver of humans and rats following chronic alcohol ingestion. Alcohol appears to increase the urinary excretion of zinc, perhaps because of increased release of zinc from hepatic stores. Protein catabolism is also associated with increased urinary zinc losses. The effect of zinc deficiency on alcohol metabolism has not been established, but preliminary evidence suggests that zinc-dependent enzymes, such as those involved in vitamin A metabolism, may be inhibited by alcohol.

When we think of antioxidants, we think of vitamins C, E, A (beta-carotene), and selenium. But, did you know the essential trace mineral, zinc, is also an important antioxidant?

We have known for some time of zinc's importance in respiration, digestion, growth and development, nerve, brain, immune system, and vision function, but only recently has zinc's antioxidant role in the body been explained. Antioxidants are the front line of defense against free radicals. They have the ability to neutralize free radicals by giving up an electron of their own without becoming harmful themselves, thus putting an end to the destructive chain reaction.

Zinc's action as an antioxidant works in many ways. First is zinc's vital role in the antioxidant enzyme *superoxide dismutase* (SOD). This is probably its most important activity as an antioxidant. SOD is a primary defender against free radicals and is so important to survival that it is the fifth most prevalent protein (of more than 100,000 in the body). SOD eliminates destructive superoxide molecules, a common free radical produced in the body. What's more, a 1992 study indicates that SOD apparently blocks the oxidation of harmful LDL cholesterol, thereby inhibiting the initial stages of atherosclerosis.[8] This offers very promising hope for combating cardiovascular disease.

Second, zinc appears to protect against free radical damage by defending sulfhydryl groups against oxidation. In the body, sulfhydryl groups are a common part of many molecules and are easily oxidized, forming free radicals.

Third, zinc limits free-radical production in the body. For example, liver cells produce a free radical known as malondialdehyde (MDA), while human neutrophils (a type of white blood cell) produce superoxide. These are both decreased by zinc.

Fourth, zinc fights free radicals by competing with prooxidant metals (iron and copper) for cell binding sites. This decreases the possibility of free-radical formation.

The Recommended Dietary Allowance (RDA) for zinc is 15 milligrams per day for men and 12 milligrams per day for women. However, factors such as diet, climate, age, stress, pregnancy, lactation, infection and level of physical activity can all affect individual requirements. If you are following a "healthy" diet—eating more fiber, more legumes, vegetables, and whole grains—you might *not* be getting enough zinc. These foods contains *oxalates* (found in many fruits and vegetables) and *phytates* (from grains), which may interfere with the absorption of dietary zinc.

Other factors which might affect zinc levels could include an increased calcium intake (widely recommended to prevent osteoporosis), which inhibits zinc absorption. In addition, iron competes with zinc, so if you're taking iron supplements, you might be zinc deficient. Exercise, even moderate physical activity (two to four hours per week) can significantly lower plasma zinc levels. And to top it off, a substantial amount of zinc is lost in perspiration.

One recent study showed that zinc deficiency, coupled with moderate alcohol intake, produced high lipid peroxidation and impaired contraction of heart tissue. Zinc deficiency has also been linked to a reduction in plasma levels of vitamins A, E, and beta-carotene. Further, the tissues of animals deficient in zinc show an increase in free-radical production.

Possible symptoms of zinc deficiency (in addition to free radical damage), which disappear with adequate zinc intake include:

1. Growth retardation
2. Skin rashes and lesions
3. Impaired taste and smell
4. Delayed wound healing
5. Immune deficiencies
6. Delayed sexual maturation
7. Night blindness
8. Low sperm count
9. Alopecia (hair loss)
10. Impaired reproduction

Sources of Zinc

Now that we've established your need for zinc, where do you get it? Traditionally, foods have been the main sources. Unfortunately, most of the foods high in zinc (such as red meat, organ meats, oysters, and nuts) are also high in fat and cholesterol. Most weight-loss programs are very low in red meat. If you are trying to limit your fat intake, you are probably also limiting the above foods, and not getting enough dietary zinc.

In addition, with the magnitude of food additives, food preservatives, radiation, long-term storage, and so forth, the zinc content of your foods might be grossly reduced.

If you are vegetarian and don't eat red meat, or if you are trying to increase your fiber intake (which interferes with the absorption of zinc), if you are under stress, or are an athlete, it may not be possible to get the needed zinc through your diet. Almost 90 percent of the body's zinc is concentrated in the bones and muscles, and therefore not readily available for other bodily functions. For all these reasons, daily supplementation seems necessary for survival.

But what form of supplements? Zinc supplements are available as zinc sulfate, zinc oxide, and in chelated forms, including zinc citrate, zinc gluconate, and now, zinc monomethionine.

Zinc sulfate and oxide, however, are poorly absorbed. Chelated forms are the most easily absorbed and utilized by the body. According to Robert M. Hackman, Ph.D., of the University of Oregon, "New product introductions which include zinc monomethionine may be particularly attractive since the zinc is well-absorbed and appears to be effectively delivered to the immune system."

Studies have shown that zinc monomethionine resists dietary fiber, phytate and oxalate. It also offers the additional benefit of providing the amino acid methionine, another potent antioxidant. Methionine is the most readily absorbed of all the amino acids.

Every day, ongoing research reveals something new about zinc and its role in human nutrition. Zinc's value as an antioxidant and free-radical scavenger underscores its tremendous importance.

So stick with zinc, the unsung antioxidant, to help you:

fight those dreaded free radicals
heal faster
get more enjoyment from the taste of food
have healthier skin and hair
fortify your immune system
reduce the tendency for atherosclerosis
live a healthier, happier, more productive life

Magnesium is involved in more than 300 enzyme systems. Necessary for energy metabolism, magnesium is required for glucose metabolism, synthesis of fats, proteins, and nucleic acid, muscle contraction, and blood clotting. In fact, magnesium is required for metabolism.

Because magnesium is necessary for certain thiamine-dependent enzymes, there has been speculation that magnesium plays a role in the development of some of the neurologic complications of alcohol abuse.

The RDA for magnesium is 280 milligrams per day for women and 350 milligrams a day for men. Good food sources of magnesium include dark green vegetables, whole grains, nuts, and seafood.

Iron is required not only for hemoglobin biosynthesis, but is also an essential component of many enzymes required for the production of energy in cells throughout the body. In iron deficiency, the impairment of these and other iron-dependent processes may account for disorders in immune function and behavior that are not directly attributable to the anemia.

Dietary iron deficiency is responsible for the most prevalent form of anemia in the United States. Iron deficiency hampers the body's ability to produce hemoglobin, a substance needed to carry oxygen in the blood. A principal consequence of iron deficiency is reduced work capacity, although depressed immune function, changes in behavior, and impaired intellectual performance may also result.

Iron deficiency has been associated with increased incidence of certain infections, perhaps because it impairs the function of peroxidase enzymes and leads to deficits in the production of free oxygen radicals and hydrogen peroxide that can kill ingested bacteria.[9,10]

The total amount of iron in the body is slightly less than the weight of a nickel, about 4 grams for adult males. Most of this iron is used to transport and use oxygen in the production of cellular energy. An average of about 58 percent of the body's iron for men and 73 percent for women is contained in hemoglobin, while 11 percent for men and 14 percent for women is present in myoglobin (which stores oxygen and makes it available to the muscle when it is needed for contraction). A smaller, but critically important amount of iron, about 3 percent, is present in *iron-containing enzymes* such as cytochromes, which are required for the production of cell energy.[11]

Copper serves as a constituent of many enzymes. Though these enzymes have many different metabolic roles, they all have one thing in common: all of them involve reactions that consume oxygen or oxygen radicals. The copper-containing enzyme lysyl oxidase helps synthesize connective tissues, while other copper-containing enzymes, ceruloplasmin and ferroxidase II, participate in the oxidation of ferrous iron to ferric iron. Probably the best known copper-containing enzyme is SOD (superoxide dismutase), which also contains zinc. SOD works as an antioxidant and protects cell membranes by eliminating free radicals.

Many enzymes and proteins (including superoxide dismutase, amine oxidase, cytochrome oxidase, uricase, ascorbic acid oxidase, tyrosinase, ceruloplasmin, and lysyl oxidase) need copper as an essential component.

In laboratory animals, symptoms of copper deficiency include enlarged hearts, anemia, defects in skeleton and vasculature related to faulty cross-linking of collagen and elastin, disorders of the central nervous system, and impaired immune function.

Manganese metalloenzymes assist in urea synthesis and the prevention of lipid peroxidation by free radicals. In addition, manganese works as a cofactor for several enzymes involved in various metabolic processes, but the number of these metalloenzymes is limited. Pyruvate carboxylase and SOD enzymes (located primarily in mitochondria) are two of the manganese-containing enzymes. However, a large number of metal enzyme complexes involving transferase, hydrolase, lyase, isomerase, and lipase reactions can be activated by manganese. It should be no surprise that symptoms of manganese deficiency are numerous, including bone abnormalities, carbohydrate and lipid disturbances, growth defects, central nervous system manifestations, and reproductive dysfunction.

Selenium is part of the enzyme *glutathione peroxidase*, and works with vitamin E in its role as an antioxidant. Rich food sources include seafood, organ meats, meat, and sometimes, whole grains (depending on the soil in which they were grown).

Low selenium intake and its resulting poor nutritional status, along with low tissue levels of glutathione peroxidase (an enzyme containing selenium) may lead to severe chronic conditions, such as cancer and cardiovascular diseases.

Molybdenum has a role in several metalloenzymes and also in iron metabolism. Because the necessary amounts of this mineral are so minute, deficiencies in humans are unknown. Even though molybdenum is required for the action of several enzymes, an excess can actually cause enzyme inhibition. Foods rich in molybdenum include whole grains, leafy green vegetables, liver, and milk.

Coenergizers in Review

Can you get all the vitamins, minerals, and enzymes you need from your foods? Those who are opposed to supplementation (and want it regulated) say that you can.

But what are your eating habits? Do you eat three square meals a day with lots of fresh fruits and vegetables? Or do you eat on the run and consume too much protein and fat? Are you on medications that could affect the proper digestion and absorption of your foods? Do you have a chronic disease? Are you an athlete, and therefore, require more nutrients than you might be able to get from your foods? Were your foods grown with pesticides, on artificially fertilized ground?

Supplements may be the only way that many of us can get all the necessary nutrients—nutrients that are critical to proper digestion and absorption.

References

1. S. Shaw and C. S. Lieber, "Alcoholism," *Nutritional Support of Medical Practice*, 2nd ed., Editors H. A. Schneider, C. E. Anderson, and D. B. Coursin. Philadelphia: Harper & Row, 1983, 236–259.
2. G. E. McKeown-Eyssen, C. Holloway, V. Jazmaji, et al., "A Randomized Trial of Vitamin C and E Supplementation in the Prevention of Recurrence of Colorectal Polyps," presented at the Eleventh Annual Meeting of the American Society of Preventive Oncology, March 11–13, 1987, San Francisco.
3. Eleanor N. Whitney and Eva M. N. Hamilton, *Understanding Nutrition*. St. Paul: West Publishing Co., 1987, 292–293.
4. L. W. Sullivan and V. Herbert, "Suppression of Hematopoiesis by Ethanol," *Journal of Clinical Investigation* 43:2048–62 (1964).

5. J. Lindenbaum, "Metabolic Effects of Alcohol on the Blood and Bone Marrow," *Metabolic Aspects of Alcoholism*, ed. C. S. Lieber. Lancaster, England: MTP, 1977, 215–47.
6. E. G. Bliznakov and G. L. Hunt, *The Miracle Nutrient: Coenzyme Q10*. New York: Bantam Books, 1987.
7. Stephen Langer, "Coenzyme Q10: For Heart and Artery Disorders," *Better Nutrition for Today's Living* 22:22 (Feb 1991).
8. H. A. Lehr, M. Becker, et al., "Super-Oxide Dependent Stimulation of Leukocyte Adhesion by Oxidatively Modified LDL *in vivo*," *Arteriosclerosis and Thrombosis* 12(7):824–829 (1992).
9. W. R. Beisel, "Single Nutrients and Immunity," *American Journal of Clinical Nutrition* 35(Feb. Supplement):417–68 (1982).
10. D. J. Stinnett, *Nutrition and the Immune Response*. Boca Raton, Fla.: CRC Press, 1983.
11. *The Surgeon General's Report on Nutrition and Health*. Washington, D.C.: U.S. Department of Health and Human Services, 1988, 470.

CHAPTER 7

How to Get Enzymes in Your Life

SO, HOW can we get more enzymes into our lives? How can we jump-start our bodies? We get more enzymes in our diet by knowing where the most active enzymes are and then eating these foods. It's like prospecting for gold! Try a two-pronged attack:

1. Diet, emphasizing raw fruits, vegetables, and complex carbohydrates
2. Supplements—enzyme therapy

IMPROVE YOUR DIET

As we've already mentioned, every plant and animal contains enzymes. Eating any animal or plant food will bring active enzymes into your body. However, because heat kills enzymatic activity, cooked or heated foods may be enzyme dead, and enzymatic activity may be diminished when food is frozen or dried. Therefore, raw foods will provide the most active enzymes. Does this mean you should eat raw beef, chicken, or fish? Well, only if you want to. Risk from parasites, bacteria (such as *E. coli*), or other goblins might dissuade you from following that course; and we don't advocate eating raw meats.

Don't forget that without a properly functioning body, even the proper type and quantity of enzymes won't be enough. Because enzymes need coenzymes and cofactors, if you're deficient in vitamins or minerals, your enzymes won't be able to do their jobs.

Fresh fruits and vegetables are the best sources of fresh enzymes. Although *all* fresh foods contain enzymes, we would like to mention some of the better known enzymes and the foods in which they naturally occur. You probably are already aware of some of the richer enzyme sources, such as papayas and pineapples.

Most of us have tasted the juicy sweetness of ripe papaya. A tropical fruit, papaya is available fresh and canned. The proteolytic enzyme papain is found in unripe papayas and in papaya leaves. Papain, commonly used in this country as a meat tenderizer, has an abundance of uses: it can clot milk, combat digestive disorders, and is also used to remove hair from hides during tanning. Almost 80 percent of all beer made in America is treated with papain, as it helps keep the beer clear.

Pineapples are one of the most popular tropical fruits. Available year-round, you should buy and consume them fresh. The heat of canning inactivates the pineapple's enzymes. Have you ever made gelatin? If so, you'll know that most packages warn you to use only canned, not fresh pineapple. This is because the protein-digesting enzymes in the fresh pineapple will break down the gelatin's protein, and it will never gel. Adding fresh pineapple to anything containing any type of protein will cause changes to occur.

Because fresh pineapples have the enzymes necessary to break down proteins, a pineapple marinade would be effective to tenderize meat before cooking. Bromelain, the enzyme in pineapple, is used in meat tenderizers, in some cosmetics, in the manufacture of beer, as well as in preparations to treat edema and inflammation.

In addition to papayas and pineapples, a number of fruits and vegetables have shown high protease activity, including guava, figs, gingerroot, kiwi, asparagus, and the Chinese gooseberry. Even some cucurbits contain high levels of proteolytic enzymes.[1]

Barley, wheat, rice bran, green beans, tomatoes, and oranges are all high in proteolytic enzymes,[2] as are corn, oats, rye, sorghum, alfalfa, mung beans, peas, peanuts, soybeans, pumpkin, spinach, and watermelon.[3] In fact, every fruit and vegetable has enzymatic activity.

Do some vegetables cause you to suffer from bloating and flatulence? This is usually because we lack the enzyme *alpha-galactosidase* which breaks down the sugars (oligosaccharides) present in beans, cabbage, broccoli, and other vegetables. When not properly broken down, these sugars stay in your digestive tract, fermenting, and causing gas. There are now products on the market, however, that supply the missing enzyme. Drops of the liquid are added to the first bites of your food. The enzyme breaks down the indigestible sugars, so gassiness and bloating are avoided.

How about nuts and seeds? They're loaded with enzymes, vitamins, and minerals. Be aware, though, that nuts, seeds, and legumes, as well as other foods including soybeans, contain "enzyme inhibitors." These are naturally occurring compounds which actually inhibit the activity of their enzymes. Inhibitors serve to keep the food from deteriorating until it can germinate (usually in the presence of moisture). Seeds, nuts, and grains are very high in enzymes, but would go bad quickly if it weren't for these inhibitors. So these foods *must* be treated to eliminate the inhibitors. Cooking will kill the inhibitors—but, unfortunately, it will also kill the enzymes in the foods. Sprouting is a good way to kill most of the inhibitors, while keeping the enzymes alive.

Another way to get more enzymatic bang for your buck is to augment your diet with food extracts from aged garlic and edible plants (such as young barley leaves and wheat grass, even certain seaweeds or algae). In fact, the renowned Japanese researcher Yoshihide Hagiwara, M.D., calls the essence of green barley, made from young barley leaves, "nature's ideal fast food."

Over twenty enzymes can be found in green barley essence, including catalase, superoxide dismutase, transhydrogenase, peroxidase and fatty acid oxidase.

And virtually every month something new is announced about the age-old "stinking rose"—garlic—particularly its usefulness in treating and preventing heart disease, stroke, cancer, as a natural antibiotic and immune-system builder, and more. Garlic is loaded with natural enzymes as well.

Get the most from your foods and enzymes. Be careful how you combine your foods. You could be blocking your active enzymes.

Beware of Processed Foods

Welcome to the "civilized" world where food processing (canning, cooking, "nuking," freezing, heating, and storage) is a way of life.

Irradiation and genetic engineering are no longer on the horizon—they're happening now!

Enzymes from "fresh" foods might have been enough for our forefathers. But in today's fast-paced, stress-filled society, truck farms (where fruits and vegetables were picked the same day they were eaten) are only fond memories. They went out with the tintype and Henry Ford's Model T. Today, "fresh" fruits and vegetables are shipped by trucks, trains, or planes, hundreds and thousands of miles before being consumed by you and me. In addition, farmers apply artificial fertilizers (phosphates) as well as pesticides, herbicides, fungicides and other "-cides" that were unknown and unused by our forefathers. If this isn't bad enough, today's food is no longer cooked fresh in your mother's kitchen, it's canned, frozen, dried, processed, reconstituted, deep-fried and probably stripped of all nutrients—including enzymes—before it gets to your mouth.

For centuries, man has milled and polished grains and legumes to make them easier to cook and digest. But did you know that modern-day mechanical milling of grains almost totally eliminates the bran and germ? After milling, wheat, rye, and corn lose about 75 percent of their mineral elements and electrolytes, resulting in losses of 16 percent for selenium, 40 percent for chromium, 48 percent of molybdenum, 68 percent of copper, 75 percent of iron, 78 percent of zinc, and 85 percent of manganese.[4] No wonder bakeries feel the need to "enrich" white bread!

If our body has a limited enzyme-producing capacity, an enzyme "bank account," as Dr. Edward Howell (in his book *Enzyme Nutrition, The Food Enzyme Concept*) has hypothesized, anything that overtaxes our digestive enzyme system, could decrease this bank. The result could be problems in digestion, toxicity, and a whole array of serious illnesses.

Temperature's Effect on Enzymes

How about cooking? Did you know that most enzymes are inactivated at 60° C (approximately 140° F)?[5] High temperatures can denature enzymes and destroy their activity. The optimum temperature for any enzyme is the temperature at which the enzymatic reaction progresses most rapidly, and this varies widely depending on the enzyme in question. For instance, the enzymes in the human body develop their highest activity at about body temperature.

When we are ill, our bodies raise temperature (with the aid of enzymes) until the enzymes are in a "fever" of activity. The fever enhances the activity of those enzymes needed to combat a health crisis. However, if our temperature is too high, enzymes are destroyed, which eventually can cause a collapse of the enzyme systems, leading to death.

On the other hand, a decrease in temperature will reduce enzyme activity. This is why food stays fresh in the refrigerator. A reduction in temperature suspends enzymatic activity, keeping food fresh, longer.

What about pasteurization? Critical in the dairy and beverage industries, there are primarily three types of pasteurization methods (see Table 7-1).[5] Because of their high heat, all of these methods will *kill* any enzymes found in the foods.

Table 7–1. Pasteurization Methods

1. Low temperature, long time (LTLT)—30 minutes at 145° F (63° C).
2. High temperature, short time (HTST). This is the most frequently used method and requires 161° F (72° C) for 15 seconds.
3. Ultrahigh temperature (UHT) 280° F (138° C) for 2 seconds.

Even simply boiling your foods will not only kill all the enzymes but will also deplete other nutrients. Navy beans, cooked in boiling water, lose between 50 percent and 65 percent of their calcium, magnesium, copper, iron, zinc, phosphorus, and electrolytes.[6]

Many store-bought cookies, ready-to-eat cereals, and other preshaped products are made through a process known as *extrusion* cooking. This process combines high pressure with moderately high temperature (usually 120° to 140° C). Though this process may be great for making Cheerios®, unfortunately, it also causes *a loss of phytase activity*. Phytase is the enzyme present in whole grains that hydrolyzes (breaks down) phytate. You may be aware that phytate (found in the husks of legumes, grains, and seeds) can tie up minerals such as iron, zinc, magnesium, calcium, and copper, keeping them from being absorbed and utilized. So if you reduce phytase, this means that phytate won't be broken down and will interfere with your absorption of these minerals.

In addition to loss of enzyme activity, researchers in one study found that, in extrusion cooking, 41 percent of the zinc was lost, 31 percent of the magnesium, and 11 percent of the phosphorus (to name a few).[5] Think of what a zinc, magnesium, or other mineral deficiency can do to your body and to the enzyme systems that depend on these nutrients.

Canning: The Enzyme Death House

Canning, because it involves heating food to the boiling point for extended periods of time, kills *all* of the enzymes in your food. In addition, many essential elements are lost. For example, in canning spinach, 80 percent of manganese is lost, 70 percent of cobalt and 40 percent of the zinc. Canned beans lose 60 percent of their zinc, while canned tomatoes lose 80 percent of this element.[5]

The high temperatures of canning can also destroy many vitamins. Canned foods may lose as much as 91 percent of B_6 and pantothenic acid.[7] In addition, canned foods are basically "enzyme dead" because most enzymes are killed at temperatures well below boiling (212° F).

Buy fresh fruits and vegetables, instead of canned ones, whenever possible.

Freezing

Freezing is probably the least damaging method of preserving food, but nothing's as good as fresh.

Most foods sold frozen are blanched, prior to freezing, in order to inactivate natural enzymes which may otherwise be still slowly active even at low temperatures, and would alter the taste or texture of the food.[5] Besides killing enzymes, this blanching can deplete vitamin C and thiamine (and other B vitamins).

Even though sun or air drying (an ancient practice for food preservation) can reduce bacterial activity and preserve food, it also arrests enzymatic processes.

Our Care Packages: Supplemental Enzymes

Because the nutrient levels of our foods are depleted by pesticides, growing conditions, processing, storage, and cooking, many people take enzyme supplements as vitamin and mineral boosters. But remember, supplements are just that, supplements, and should be taken in addition to a well-balanced, daily diet.

Even if we're getting the most optimal diet possible, we might need enzyme supplements if we have a problem digesting or absorbing the nutrients we eat. As Jeffrey Bland, Ph.D. said, "You're not what you eat, but what you absorb."

What if you are eating correctly, digesting properly, and also absorbing the nutrients? That's great, *unless* your immune system is depleted, or you've got a buildup of cholesterol (which leads to heart problems), or you have AIDS, herpes, cancer, and so forth. Any of these conditions (and others) can severely impact your healthy status, and therefore, your enzyme systems. Supplemental enzymes might be required.

Although there are many reasons for taking enzymes (and we'll cover some of them later), most people take them as digestive aids. In addition, the substitution of enzymes in intestinal deficiency is a classic treatment.

You can buy enzymes in the health-food store and even in most grocery stores. Available in tablets, capsules, or powder, most enzymes are taken orally. However, some enzyme products can be taken sublingually (under the tongue), topically (in ointment form), rectally, or even by injection (in a hospital setting).

How much should you take? Because we're all different, and because enzyme products vary in strength and composition, check the label on the bottle for dosage information.

Enzymes are usually well-tolerated, though you might notice stool softening, flatulence, or a feeling of fullness. If so, reduce the dosage.

For a long time, people didn't think we could absorb supplemental enzymes. We now know that we can absorb enzymes in a number of ways, primarily through a mechanism known as *pinocytosis*. Pinocytosis is actually a system whereby enzymes, after connection to a receptor in the mucosa of the intestinal wall, are absorbed into that wall, guided through the intestinal cells, and finally released into the blood—much like an elevator taking you from one floor to the next.

Some enzymes (depending on the type) are enterically coated to dissolve in the small intestine, instead of the stomach.

Enzymes are made from a variety of sources, including hog and bovine pancreas, pineapple and papaya fruits, and Aspergillus molds.

Some of the enzyme preparations sold in health-food stores, drug and grocery stores, and through multilevel marketing and mail order contain papain, pancreatin, trypsin, chymotrypsin, bromelain, amylase, lipase, or lactase. Others may also contain cellulase, brinase, serratiopeptidase, SOD (superoxide dismutase), or diastase, among others.

The activities of enzymes most frequently used are proteolytic (protein-lysing), lipolytic (fat-lysing), and amylolytic (amylase-lysing), as shown in Table 7–2.

Table 7–2. Enzymatic Activities

ENZYMES	TYPE OF ACTIVITY
Pancreatin	Proteolytic activity Lipolytic activity Amylolytic activity
Trypsin	Proteolytic activity
Chymotrypsin	Proteolytic activity
Papain	Proteolytic activity
Bromelain	Proteolytic activity
Amylase	Amylolytic activity
Lipase	Lipolytic activity
Lactase	Amylolytic activity
Cellulase	Amylolytic activity

CHARACTERISTICS OF JUMP-START ENZYMES

Pancreatin

Source: animal

Is unique because it possesses proteolytic, lipolytic, and amylolytic activities

Contains proteolytic enzymes (chymotrypsins, trypsins, pancreato- and carboxypeptidases), amylases, lipases, phospholipases, as well as nucleic acids (RNA, DNA)
Most active from pH 6.5 to 9.0

Uses:
In pancreas insufficiency, inadequate secretion of exocrine pancreas, disturbed digestion and after gastrectomy[8–21]
In children with cystic fibrosis[22–26]

Trypsin

Source: animal (ox pancreas)[27]
Has endoproteolytic properties and splits peptides, amides, and esters[28]
Activities are optimal at pH 7.0 to 9.0

Uses:
In debridement of necrotising wounds, ulcerations, abscesses, empyemas, hematomas, fistulas, and decubitus ulcers, trypsin is used locally and externally[29–32]
To accelerate healing in injuries, inflammations, phlogistic edemas and traumatic changes[33–40]
Auxiliary agent in meningitis therapy[41]

Chymotrypsin

Source: animal (ox and pork pancreas)[42]
Has endoproteolytic activities and splits peptides, amides, esters and other amino acid-containing compounds[42]
Optimal efficiency at pH 8.0

Uses:
In debridement, in treatment of abscesses and ulcerations, in liquefaction of mucous secretions[43]
In ophthalmic cataract surgeries and in therapy of eyeball hematomas and ophthalmorrhagias[43,44]
Before and after tooth extractions as well as in operative dentistry[45,46]

After episiotomy surgeries[47]
As an anthelmintic against enterozoic worms[48]
In early recognition of tumor cells[49]
In histologic gastroenterologic diagnostics[50]

Papain

Source: Carica papaya *(papaya latex)*[27]
Splits peptides, amides and esters[51]
Optimal enzymatic activity from pH 2.5 to 7.0

Uses:
In ophthalmology to prevent cornea scar malformation[52]
In intoxications caused by stings of jellyfish and insects[53]
In malabsorption syndrome caused by gluten intolerance[54]
In treating phlogistic edemas, inflammatory processes and in the acceleration of wound healing[55–63]

Bromelain

Source: Ananas comosus *(pineapple stem)*
Is a mixture of bromelain A and B. These two sulfur-containing proteinases split peptides, amides, as well as glycine esters[51]
Activity is highest between pH 3.0 to 8.0

Uses:
As an adjuvant in treatment of swelling and inflammations due to injury and surgery[64–65]
In painful menstrual hemorrhages[66]
In boxing injuries, such as facial swellings and hematomas[67]
In acute sinusitis[68]
In thromboembolism of central retinal vessels[69]

Amylase

Source: fungal (Aspergillus oryzae)[27]
Needs calcium ions for enzymatic activity

Acts on starch, glycogen and related poly- and oligosaccharides[70]
Optimal efficiency at pH 6.5
Is used, in combination with other enzymes, as a digestant[43,71,72]

Lipase

*Source: fungal (*Aspergillus oryzae*)*
Activity depends on calcium ions
Splits emulsified neutral fats into fatty acids and glycerol
Highest hydrolytic activity from pH 5.0 to 7.5

Uses:
To increase pancreatic/lipolytic activities in pancreatin-containing remedies[73–76]
Reduces fat level in stools when given in combined preparations with pancreatin[77–79]
Intensifies synergistically biocatalytic activity of lipoprotein-lipase in blood[80] and migration of agranulocytes[81]

Lactase

Source: Aspergillus niger *(fungus)* and Saccharomyces lactis *(yeast)*[82]
Optimal pH varies 4.0 to 5.0 pH from *A. niger* sources; 6–8.5 pH from *S. lactis* sources[82]
Used to treat lactose insufficiency and as a digestive aid

Cellulase

*Source: fungal (*Aspergillus niger*)*
Acts to break down cellulose and cereal glucans[70]
Used as a digestive aid

Enzymes are used individually and in combinations. Some people feel that combinations are more effective than individual enzymes because of their synergistic effect. However, this is an individual preference.

Table 7–3. Optimal Enzyme pH

	pH RANGE
Animal Origin	6.5–9.0
Pancreatin	
Trypsin	7.0–9.0
Chymotrypsin	8.0
Plant Origin	
Papain	2.5–7.0
Bromelain	3.0–8.0
Fungal Origin	
Amylase	6.6
Lipase	5.0–7.5
Lactase	4.0–5.0 (if from *Aspergillus niger*)
	6.0–8.5 (if from *Saccharomyces lactis*)

Enzymes at a Glance

We have a multitude of jump-start enzymes available to us in a cornucopia of fresh fruits, vegetables, meats and dairy products, plus concentrated supplement forms. Nature has given us the tools for healthier, happier lives. It is now time for us to wake up and make our dreams come true through a simple 5-Step Jump-Start Plus Enzyme Program.

References

1. Michael J. Dalling, Editor. *Plant Proteolytic Enzymes*. Boca Raton, Fla: CRC Press, Inc., 1986, 81–86.
2. Ibid., 99–101
3. Ibid., 124–134.
4. R. O. Nesheim, "Nutrient Changes in Food Processing. A Current Review," *Federal Procedures* 33:2267 (1974).
5. R. A. Wapnir, *Protein Nutrition and Mineral Absorption*. Boca Raton, Fla: CRC Press, 1990, 62–66.

6. C. R. Meiner, N. L. Derise, et al., "Proximate Composition and Yield of Raw and Cooked Mature Dry Legumes," *Journal of Agricultural and Food Chemistry* 24:1122 (1976).
7. Jane Brody, *Jane Brody's Nutrition Book*. Toronto: Bantam Books, 1987, 178.
8. E. LeBauer, K. Smith, and N. J. Greenberger, "Pancreatic Influence and Vitamin B_{12} Malabsorption," *Archives of Internal Medicine* 122:423–426 (1968).
9. M. S. Kataria and D. Bhaskarrao, "A Clinical Double-Blind Trial with a Broad Spectrum Digestive Enzyme Product (Combinzym) in Geriatric Practice," *British Journal of Clinical Practice* 23(1):15–17 (1969).
10. S. Karani, M. S. Kataria, and A. E. Barber, "A Double-Blind Clinical Trial With a Digestive Enzyme Product." *British Journal of Clinical Practice* 25:375–377 (1971).
11. R. P. Knill-Jones, H. Pearce, H. Batten, and R. Williams, "Comparative Trial of Nutrizym in Chronic Pancreatic Insufficiency," *British Medical Journal* 4:21–24 (1970).
12. E. P. DiMagno, V. L. W. Go and W. H. J. Summerskill, "Relations Between Pancreatic Enzyme Outputs and Malabsorption in Severe Pancreatic Insufficiency," *New England Journal of Medicine* 288:813–815 (1973).
13. J. H. B. Saunders and K. G. Wormsley, "Progress Report: Pancreatic Extracts in the Treatment of Pancreatic Exocrine Insufficiency," *Gut* 16:157–162 (1975).
14. S. Bank, I. N. Marks, and G. O. Barbezat, "Treatment of Acute and Chronic Pancreatitis," *Drugs* 13:373 (1977).
15. E. P. DiMagno, J. R. Malagelada, V. L. W. Go, and C. G. Moertel, "Fate of Orally Ingested Enzymes in Pancreatic Insufficiency: Comparison of Two Dosage Schedules," *New England Journal of Medicine* 296:1318–1322 (1977).
16. J. H. Meyer, "The Ins and Outs of Oral Pancreatic Enzymes," *New England Journal of Medicine* 296:1347–1348 (1977).
17. J. H. B. Saunders, S. Drummond, and K. G. Wormsley, "Inhibition of Gastric Secretion in Treatment of Pancreatic Insufficiency," *British Medical Journal* 1:418–419 (1977).
18. Anonymous, "Pancreatic Extracts," *Lancet* 2:73–75 (1977).
19. P. T. Regan, J. R. Malagelada, et al., "Comparative Effects of Antacids, Cimetidine and Enteric Coating on the Therapeutic Response to Oral Enzymes in Severe Pancreatic Insufficiency," *New England Journal of Medicine* 297:854–858 (1977).
20. P. T. Regan, J. R. Malagelada, E. P. Dimagno, and V. L. Go, "Rationale for the Use of Cimetidine in Pancreatic Insufficiency," *Mayo Clinic Procedures* 53:79–88 (1978).

21. W. J. Austad, "Pancreatitis; The Use of Pancreatic Supplements," *Drugs* 17:480–487 (1979).
22. Anonymous, "Pancreatic Extracts," *British Medical Journal* 2:161–163 (1970).
23. C. M. Anderson, "Pancreatic Enzyme Replacement in the Treatment of Cystic Fibrosis," *Prescribers Journal* 12:45–49 (1972).
24. C. C. Roy, A. M. Weber, et al., "Abnormal Biliary Lipid Composition in Cystic Fibrosis," *New England Journal of Medicine* 197:1301–1305 (1977).
25. A. C. Smalley, G. A. Brown, et al., "Reduction of Bile Acid Loss in Cystic Fibrosis by Dietary Means," *Archives of Disease in Childhood* 53:477–482 (1978).
26. M. C. Goodschild, E. Sagaro, et al., "Comparative Trial of Pancrex V. Forte and Nutrizym in Treatment of Malabsorption in Cystic Fibrosis," *British Medical Journal* 3:712–714 (1974).
27. Malcom Dixon, Edwin C. Webb, C. J. R. Thorne, and K. F. Tipton, *Enzymes*. New York: Academic Press, 1979, 550–551.
28. Ibid., 887.
29. Heinrich Wrba, "Trypsin aund Chymotrypsin-Klinisches Sachverstan-Digengutachten zur Systemischen Wirkung," Inst. f. angewante u. exptl. Onkologie d. Univ. Wien. (Nov. 1988).
30. G. Stille and K. Tuluwett, "Pharmakologisch-Toxikologisches Sach-Verstandigengutachten zut Trypsin/Chymotrypsin," Nov. 1988.
31. B. Gordon, "The Use of Topical Proteolytic Enzymes in the Treatment of Post-Thrombotic Leg Ulcers," *British Journal of Clinical Practice* 29:143–146 (1975).
32. M. R. Sather, C. E. Weber, Jr., and J. George, "Pressure Sores and the Spinal Cord Injury Patient," *Drug Intelligence & Clinical Pharmacy* 11:154–169 (1977).
33. Samuel E. Soule, Helman C. Wasserman, Robert Burstein, "Oral Proteolytic Enzyme Therapy (Chymoral) in Episiotomy Patients," *American Journal of Obstetrics and Gynecology*, 95(S):820–833 (1966).
34. P. S. Boyne and H. Medhurst, "Oral Anti-Inflammatory Enzyme Therapy in Injuries in Professional Footballers," *Practitioner* 198(S):543–546 (1967).
35. J. L. Blonstein, "Oral Enzyme Tablets in the Treatment of Boxing Injuries," *Practitioner* 198(S):547–548 (1967).
36. J. E. Buck and N. Phillips, "Trial of Chymoral in Professional Footballers," *British Journal Clinical Practice* 24:375–377 (1970).
37. W. F. Rathgeber "The Use of Proteolytic Enzymes (Chymoral) in Sporting Injuries," *South African Medical Journal* 45(S):181–183 (1971).

38. A. De N'Yeurt, "The Use of Chymoral in Vasectomy," *Journal of the Royal College of General Practitioners* 22:633–637 (1972).
39. Travis Winsor, "Inhibition of the Response to Thermal Injury by Oral Proteolytic Enzymes," *Journal of Clinical Pharmacology* 12(S):325–330 (1972).
40. W. F. Rathgeber, "The Use of Proteolytic Enzymes in Tenosynovitis," *Clinical Medicine* 80(S): 39–41 (1973).
41. O. H. D. Marquez and F. G. Segur, "The Intrathecal Use of Proteolytic Enzymes in Tuberculous Meningoencephalitis," Preliminary Communication; Abstr. *World Medicine* 42:800–801 (1968).
42. *Enzymes*, 885.
43. J. E. P. Reynolds, ed., Martindale - *The Extra Pharmacopoeia*: 21. Aufl. London: The Pharmaceutical Press, 1982.
44. F. I. D. Konotey-Ahulu, "Enzyme Treatment of Vitreous Haemorrhage," *Lancet* 2(S):714–715 (1972).
45. J. H. Sowray, "An Assessment of the Value of Lyophilised Chymotrypsin in the Reduction of Post-Operative Swelling Following the Removal of Impacted Wisdom Teeth," *British Dental Journal* 110:130–133.
46. Frederick Wigand and Eugene Messer, "Enzyme Treatment of Traumatic Swelling in Oral and Maxillofacial Surgery," *Clinical Medicine* 74(S):29–31 (1967).
47. Heath D. Bumgardner and Gerald I. Zatuchni, "Prevention of Episiotomy Pain with Oral Chymotrypsin," *American Journal of Obstetrics and Gynecology* 92(4):514–517 (1965).
48. R. A. Fiel, "Tratamiento Experimental de la Trichuriasis Masiva Infantil con Quimotripsina," *Tropical Disease Bulletin* 65:917 (1968).
49. Masayoshi Takahashi, Keisuke Hashimoto, and Hiroshi Osada, "Parenteral Administration of Chymotrypsin for the Early Detection of Cancer Cells in Sputum," *Acta Cytology* 11:61–63 (1967).
50. L. L. Brandborg, C. B. Tankersley, and F. Uyeda, " 'Low' Versus 'High' Concentration Chymotrypsin in Gastric Exfoliative Cytology," *Gastroenterology* 57:500–505 (1969).
51. *Enzymes*, 890–891.
52. G. L. Starkow and W. J. Sawinych, "Papaintherapie der Augenkrankheitein. Klin. Mbl.," *Augen-Heilkunde* 159:755–759 (1971).
53. John S. Loder, "Treatment of Jellyfish Stings," *Journal of the American Medical Assoc.* 226(S):1228 (1973).
54. M. Messer and P. E. Baume, "Oral Papain in Gluten Intolerance," *Lancet* 2(S):1022 (1976).
55. P. J. Pollack, "Oral Administration of Enzymes from Carica Papaya: Report of a Double-Blind Clinical Study," *Current Therapeutic Research* 4(S):229–237 (1962).

56. P. Thorex and J. K. Pandit. "Proteolytic Enzymes in Wound Repair; Immediate Postoperative Effect," *Therapeutics* 6:323–325 (1964).
57. C. T. Yarington and J. Michael Bestler, "A Double-Blind Evaluation of Enzyme Preparation in Postoperative Patients," *Clinical Medicine* 71(S):710–712 (1964).
58. P. S. Metro and R. B. Horton, "Plant Enzymes in Oral Surgery," *Pharmacology and Therapeutics* 19:309–316 (1965).
59. Charles P. Vallis and Mark H. Lund, "Effect of Treatment with *Carica Papaya* on Resolution of Edema and Ecchymosis Following Rhinoplasty," *Current Therapeutic Research* 11(S):356–359 (1969).
60. Henry T. Holt, "*Carica Papaya* as Ancillary Therapy for Athletic Injuries," *Current Therapeutic Research* 11(10):621–624 (1969).
61. Mark H. Lund and Richard R. Royer, "*Carica Papaya* in Head and Neck Surgery," *Archives of Surgery* 98(S):180–182 (1969).
62. F. Caci and G. M. Gluck, "Double-Blind Study of Prednisolone and Papase as Inhibitors of Complications in Oral Surgery," *Journal of the American Dental Assoc.* 93:325–327 (1976).
63. W. Van Eimeren and W. Lehmacher, "Biometrische Stellungnahme zum Wirkungsnachweis von Papain," Klinikum der University Ulm. (Jan. 1989).
64. G. Stille, "Klinisches Sachverstandigen-Gutachten nach Paragraph 24 AMG fur Bromelain," paper from Mucos Pharma GmbH & Co., Geretsried, Germany, Dec. 1985.
65. W. M. Cooreman, S. Schwarpe, J. DeMeester, and A. Lauwers, "Bromelain, Biochemical and Pharmacological Properties," *Pharmaceutica Acta Helvetiae* 51:73–97 (1976).
66. S. L. B. Duncan, J. H. Lawrie, and H. R. MacLennan, "Bromelain and the Cervix Uteri," *Lancet* 2:1420–1422 (1960).
67. J. L. Blonstein, "Control of Swelling in Boxing Injuries," *Practitioner* 203 (S): 206 (1969).
68. R. Z. Ryan, "A Double-Blind Clinical Evaluation of Bromelains in the Treatment of Acute Sinusitis," *Headache* 7(S):13–17 (1967).
69. Leon F. Gray, "Thrombosis of the Central Retinal Vein and Its Treatment with Ananase," *Southern Medical Journal* 62(S):11 (1969).
70. *Enzymes*, 860–861.
71. M. Windholz, ed. *The Merck Index - An Encyclopedia of Chemicals, Drugs, and Biologicals*; 10. Rahway, New Jersey: Merck & Co. Inc., 1983.
72. H. Auterhoff and J. Knage, "Lehrouch der Pharmazueutischen Chemie," *Verlagsgesellschaft*; 11. Stuttgart Auflk, Wissenschafel, 1983.
73. D. Y. Graham, "Enzyme Replacement Therapy of Exocrine Pancreatic Insufficiency in Man," *New England Journal of Medicine* 296(S): 1314–1317 (1977).

74. J. H. Meyer, "The Ins and Outs of Oral Pancreatic Enzymes," *New England Journal of Medicine* 296(S):1347–1348 (1977).
75. T. L. Yeh and M. L. Rubin, "Potency of Pancreatic Enzyme Preparations," *New England Journal of Medicine* 297(S):615–616 (1977).
76. R. Kirshen, "Letter to the Editor," *New England Journal of Medicine* 297(S):616 (1977).
77. R. D. Mackie, A. S. Levine, and Michael D. Levitt, "Malabsorption of Starch in Pancreatic Insufficiency," *Gastroenterology* 80(S):1220 (1981).
78. P. G. Lankisch and W. Creutzfeldt, "Therapy of Exocrine and Endocrine Pancreatic Insufficiency," *Clinics in Gastroenterology* 13(S): 985–999 (1984).
79. M. U. Schneider, M. L. Knoll-Ruzicka, S. Domschke, G. Heptner, and W. Domschke, "Pancreatic Enzyme Replacement Therapy; Comparative Effects of Conventional and Enteric-Coated Micropheric Pancreatin and Acid-Stable Fungal Enzyme Preparations on Steatorrhoea in Chronic Pancreatitis," *Hepatology and Gastroenterology* 32S:97–102 (1985).
80. D. A. Hall, A. R. Zajac, R. Cox, and J. Spanswick, "The Effect of Enzyme Therapy on Plasma Lipid Levels in the Elderly," *Artherosclerosis* 43S: 209–215 (1982).
81. S. Tylewska, S. Tyski, and W. Hrynie-Wicz, "The Effect of S. Aureus Lipase on Granulocyte Chemotaxis," *Medycyna Doswiadczalna I Mikrobiologia* 35S:171–174 (1983).
82. Sigmund Schwimmer. *Source Book of Food Enzymology*. Westport, Conn.: The AVI Publishing Co., Inc., 1981, 652.

CHAPTER 8

The Five-Step Jump-Start Plus Enzyme Program

AS INHABITANTS of this great country, this world, we are in a survival mode. Our energy, our health, and our very existence are based on enzymes and the prevention of illness.

But what if we do get sick? What if enzyme production and availability is reduced? Sickness is a one-way trip downhill—emotionally, financially, as well as healthwise. Whatever you do, *don't get sick in America!* The cost of illness can devastate you and your family.

Heart disease remains the leading cause of death in the United States. Despite the fact that the death rate for coronary heart disease has declined dramatically in recent years, it remains the leading cause of death in the United States—accounting for more than 40 percent of all deaths.[1] According to the Centers for Disease Control, cardiovascular disease costs the nation approximately $286.5 billion every year.[2] These costs are largely preventable.

The five leading causes of death are heart disease, cancer, stroke, chronic obstructive pulmonary disease, and accidents, according to the U. S. government. Most of these conditions are preventable. However, the changes in lifestyle, lost productivity, and direct medical cost of these conditions are staggering.

Most of us go to the doctor only when symptoms of disease appear. We expect an instant cure and are given either drugs or surgery, the two most expensive methods available.

So what can you do to prevent illness, add years to your life, and lower your medical costs? Try detoxification, fasting, improved nutrition and diet, regular exercise, as well as nutritional supplements and a positive mental attitude to help prevent illness and injury. This will add *quality* years to our lives and jump-start each day.

Today, Americans are taking a more active interest in their health than ever before. They are learning to be responsible for their own health. They don't want to die before their time and they don't want to live lives of misery, pain, or suffering. Americans are beginning to realize the influence that they, themselves, can have on their own health. They don't want to give away their freedom.

Remember, whoever pays for health care and whatever the method of payment (whether private insurance or national government), ultimately you and I foot the bill. We either pay now (insurance premiums), or pay later (taxes). But we do pay. There are no handouts.

Therefore, the most cost-effective health plan is to *take charge of our own lives.* Better *not* to get sick. Prevent it! Let live enzymes in our bodies and in foods work for us in the Five-Step Jump-Start Plus Enzyme Program. Your days, your life, can come alive with the Five-Step Jump-Start Plus Enzyme Program.

What is the Five-Step Jump-Start Plus Enzyme Program? This program is a method of reenergizing your total body and mind. It seems so simple, but if we stay healthy, the chances of getting sick are greatly decreased. The Five-Step Jump-Start Plus Enzyme Program is our road map to good health. Follow it as we travel through the various conditions and disorders and apply it as it relates to *you.* May good health and energy be with you!

The Five-Step Jump-Start Plus Enzyme Program:

1. Detoxify (detoxification, fasting, juicing).
2. Eat a well-balanced "jump-start" diet.
3. Use enzyme, vitamin, and mineral supplements, to jump-start your day.
4. Exercise daily.
5. Have a positive mental attitude through mind and spirit power.

DETOXIFICATION, FASTING, AND JUICING

Detoxification is the removal of toxins from the body. These toxins can develop from our external environment or be produced by our body and can actually be the reason for a disease process. When toxins originate from within, it is because of the body's inability to rid itself of metabolic waste products, incomplete oxidation in the tissues, inadequate nutrition, and/or because of some existing disease process within the body. The enormous job of eliminating toxic substances places heavy demands on the liver.

Maintaining the function of the large-scale excretion mechanism of the body (the intestines) is the first step in detoxification. A great deal can be achieved by 1) restoration of the intestinal flora and 2) reestablishment of intestinal peristalsis, if necessary, with a fiber-rich diet, or (in exceptional cases) colonic irrigation.

Fasting

Fasting, that is, avoiding all food and drink (except water) has long been used religiously, as well as therapeutically. Fasting increases elimination of wastes and can augment the body's healing processes. For this reason, fasting is often used as a detoxification method. Fasting tends to rid the body of toxic products of food allergy, which is very common in today's society.

Caution is advised with fasting, for although the intense hunger of the first days may be overcome with difficulty or with ease, it is the resumption of eating that can truly be hazardous. While individuals on a regular diet may be able to handle the frequent ingestion of allergens, after a fast, the reintroduction of an allergen may cause a severe reaction. Heart rhythm disturbances have been linked, in certain cases, to food allergy.

The natural nutrients in the fresh fruits and vegetables you eat can also aid tremendously in the detoxification process. Juices are particularly beneficial, especially celery, parsley, and carrot juices. Vitamins A, C, E, and the minerals selenium and zinc, as well as SOD (superoxide dismutase) and other enzymes that you obtain from your foods, supplements, and juices are very good detoxifying agents. Not only do these nutrients have a direct effect on the toxins themselves, they also help restore overall health to organs that play a major role in detoxification.

When fasting longer than three to five days, a physician should carefully monitor the process. Many conditions, such as allergies, irritable bowel syndrome, depression, and digestive problems can often be relieved by a short fast. See Michael T. Murray and Joseph E. Pizzorno's book, *Encyclopedia of Natural Medicine* (Prima Publishing, Rocklin, Calif., 1991), for further information.

Juicing

One way to detoxify our bodies is to eat a natural diet, high in vegetables and fruits, and to drink six to eight glasses of water daily. Many of these fruits and vegetables can be combined in the form of delicious juices. Juicing should be considered an addition to the overall program. A fiber supplement (such as pectin, psyllium, or bran), vitamin C, and a high-potency vitamin/mineral/enzyme tablet can aid in the detoxification process. When juicing, the addition of fiber is important since the majority is separated away during the juicing process.

Nutrition and Diet

J. B. Cordaro, president and chief executive officer of the Council for Responsible Nutrition, an association of the dietary supplement industry, once said "Proper nutrient intake not only helps prevent disease, but is essential to maintaining good health. Unfortunately, most Americans do not get even the Recommended Dietary Allowance (RDA) of many nutrients from their normal diets. A national survey of over 21,000 people showed that not one of them received 100 percent of the RDA for ten basic nutrients from their diet alone. The National Cancer Institute recommends we get five servings of vegetables and fruit per day. On any given day, 40 percent of Americans do not have even one serving of fruit or fruit juice, and 20 percent have no servings of fruit or fruit juice over a period of four days."[4]

Dietary factors are associated with five of the ten leading causes of death, such as coronary heart disease, stroke, noninsulin-dependent diabetes mellitus, atherosclerosis, and some types of cancer.[5] In general, we should eat less fat, less protein, and more complex carbohydrates.

You may say, "I already eat right." But can you be sure? Where were the fruits and vegetables grown? What about that shiny, red apple? It didn't get that way by accident. How many times was it sprayed and with what (pesticides, herbicides, wax, etc.)? It is common knowledge that the nutritional content of food can vary widely depending on time since harvesting, growth site, storage conditions, and the distance transported from the field to you.

Man has been cooking his food for only a few thousand years. He evolved in an environment of raw fruits, vegetables and grains with little meat. His metabolism has genetically accepted or adapted to this natural course over several million years. Only within approximately the last 100 years have we begun to freeze, can, or chemically tamper with our food. Some chemicals used in food processing are toxic and may destroy enzymes and induce cancer. The cooking, canning, freezing, and preserving of foods virtually eliminates their active enzymes as well as removing many of their nutrients.

Though some foods, such as white bread, may fill us up, they have virtually no fiber or enzymes. Their only nutrients are those that were added back in after processing (to meet government guidelines).

Many of us eat too much protein and follow a "meat and potatoes" diet. Animal proteins are among the most complex foods for man to digest, but it *is* possible to obtain almost all necessary protein from vegetables.

Many of us are virtually addicted to refined carbohydrates, particularly refined sugar (in its various forms in food and drink), and we overuse salt (an indirect enzyme inhibitor).

Caffeine (in coffee, tea, and soft drinks) can cause hypoglycemia and has other negative physiological effects. When it is combined with sugars (as in cola drinks), it can be devastating to your blood sugar level.

We must change our eating habits by eliminating or at least reducing the intake of animal proteins, refined carbohydrates, and stimulants. We need to quit eating refined, canned, frozen, or otherwise processed food and drink. We must begin to rely on natural, whole fruits, vegetables and grains, in their raw, natural state, in order to obtain the most enzyme-rich foods possible.

The Five-Step Jump-Start Plus Enzyme Program Dietary Do's and Don'ts

1. Eat fresh fruits and vegetables as much as possible.
   a. Avoid foods that have been grown close to highways and those sprayed with insecticides.
   b. Thoroughly scrub fruits and vegetables.
   c. Eat fruits and vegetables whole, including the skins and seeds, except when otherwise indicated.
2. Include a lot of garlic and onions in your diet.
3. Limit foods that contain enzyme inhibitors in the diet, such as soybean products, peanuts, lentils. Instead, try sprouting.
4. Do not use aluminum cooking utensils. Use stainless steel instead. In addition, make sure any ceramic dinnerware or crystal is lead-free.
5. Avoid common table salt; some feel it is an indirect enzyme inhibitor.
6. Avoid refined sugar and products made from refined carbohydrates, such as pies, cakes, desserts, and candies.
7. Include adequate bulk in your diet by eating complex carbohydrates, whole grains, fresh fruits, and vegetables.
8. Drink large amounts of freshly made vegetable and fruit juices (such as carrot, beet, pineapple, etc.).
9. Avoid excessively cold or hot beverages or foods.
10. Eat small portions of food (five or six times a day) so as not to overtax your digestive system.
11. Avoid coffee. Drink herb teas instead.

Some people prefer to take their fruits and vegetable in juice form. For further information on juicing, see books such as *Getting the Best Out of Your Juicer* or *The Book of Raw Fruit and Vegetable Juices and Drinks,* both by William H. Lee, R.Ph., Ph.D. (Keats Publishing, Los Angeles); *The Pocket Handbook of Juice Power,* by Carlson Wade (Keats Publishing), *The Complete Book of Juicing,* by Michael T. Murray, N.D. (Prima Publishing, Rocklin, California) or *The Juicing Book,* by Steven Blauer (Avery Publishing Group, Garden City Park, New York).

We live in a society where everyone eats poorly. Many Americans are affluent, yet eat worse than their forefathers—therefore, the necessity for supplements.

NUTRITIONAL SUPPLEMENTS

Moses Maimonides (physician to the Sultan) in the twelfth century said, "No illness which can be treated by diet should be treated by any other means."[6] This is true, in theory. However, the nutrient level of today's foods is grossly depleted. It is difficult, if not impossible, to obtain the necessary nutrients in our foods. Therefore, in order to offset this deficiency, supplements are critical.

The FDA estimates that half of all adult Americans use nutritional supplements.[7] Why do we take supplements?

As you know, many vitamins and minerals are necessary for the proper functioning of our body's enzymes.

1. There is not enough time to eat correctly.
2. Due to extended storage time, processing, additives, preservatives, colors, flavors, and so forth, there is reduced nutrient value in the foods we do eat.
3. Because of today's physical, emotional, and environmental stresses there is an increased need for additional nutrients.

EXERCISE

Regular physical activity can help to manage and prevent hypertension, coronary heart disease, obesity, noninsulin-dependent diabetes, mental health problems (such as depression), and osteoporosis.[8] In addition, inactive people tend not to live as long as those who are active.[9] Further, physically active people are almost half as likely to develop coronary heart disease as people who do not engage in regular physical exercise.[10]

We feel better when we exercise because endorphin (a morphine-like brain hormone) levels increase. This produces a sense of well-being. Further, it increases circulation, brings nutrients (including enzymes) to your cells, and takes waste products away.

It is important that we exercise daily. Try doing aerobics, taking a walk, swimming, and so on, anything to increase your cardiovascular output and your oxygen intake. In this way you can jump start your energy, enzymes, and state of mind.

Positive Mental Attitude

"Attitude adjustment" and "nutrition of the mind," as Dr. Joan Priestley calls it, is the fifth step of the program. This part involves stress reduction and is accomplished through meditation, yoga, deep breathing, affirmation, visualization, forgiveness (if applicable), cassette tapes, and empowerment seminars.

"Many people carry around a lot of emotional baggage," says Dr. Priestley. "It's apparent that the greatest vitamin program and diet in the world can't begin to overcome a hateful attitude." According to Dr. Priestley, "People with disease need to resolve their feelings and reduce the stress to which they are exposed."

It is important to realize we have power over what we think and do. Dwelling on the past is not helpful. *Think positively and take charge.*

Yoga and meditation groups might be helpful, as well as reading books such as *The Power of Positive Thinking*, by the late Norman Vincent Peale, and those by Louise Hay, since mental attitude has been shown to affect the immune system.

In Review

Let the Five-Step Jump-Start Plus Enzyme Program help you to take charge of your life. Feel good, look good, live longer! If and when injured, bounce back faster! Fight chronic disease.

Apply the Five-Step Jump-Start Plus Enzyme Program to each condition in the following chapters as it fits your needs. This is your road map, your manual, for better health.

References

1. Centers for Disease Control, Chronic Disease Prevention. *Preventing Cardiovascular Disease: Addressing the Nation's Leading Killer At-A-Glance 1999*. Retrieved from http://www.cdc.gov/nccdphp/cvd/cvdaag.htm.
2. Ibid.
3. National Vital Statistics Report, Vol. 47, No. 9, Washington, D.C.: Centers for Disease Control, November 10, 1998.

4. J. B. Cordaro, Testimony to U.S. House of Representatives Committee on Energy and Commerce, Subcommittee on Health and the Environment. Hearing on legislative issues related to the regulation of dietary supplements. July 29, 1993.
5. *Healthy People 2000 National Health Promotion and Disease Prevention Objectives.* Washington, D.C.: U.S. Department of Health and Human Services, Public Health Services, 1991, 112.
6. "Diet Related to Killer Diseases, V Nutrition and Mental Health" Hearing before the Select Committee on Nutrition and Human Needs of the U.S. Senate, 95th Congress, June 22, 1977. Washington, D.C.: 1977.
7. Joseph A. Levitt, Director, Center for Food Safety and Applied Nutrition, Food and Drug Administration, Department of Health and Human Services, testimony before the Committee on Government Reform, U.S. House of Representatives, May 27, 1999.
8. S. S. Harris, C. J. Caspersen, G. H. DeFriese, and E. H. Estes, "Physical Activity Counseling for Healthy Adults as a Primary Preventive Intervention in the Clinical Setting," *Journal of the American Medical Association* 261:3590–3598.
9. A. Katz, L. G. Branch, et al., "Active Life Expectancy," *New England Journal of Medicine* 309:1218–1224 (1983).
10. K. E. Powell, C. J. Caspersen, et al., "Physical Activity and Chronic Disease," *American Journal of Clinical Nutrition* 49:999–1006 (1989).

PART THREE

The Fountain of Youth

We'd all like to look good, feel good, and live longer. Not to mention looking and feeling younger. How can we slow that roller-coaster ride called aging? There is an answer. By caring for our skin and improving our bodies through enzymes, detoxification, exercise, supplements, improved diet, and weight loss we can put the problems of aging behind us and jump-start our lives.

Chapter 9

Your Fountain of Youth

Staying Young Longer

Alice and Jane are both seventy years old. Alice fast-walks five miles every day, works out in the local gym three days a week, and attends aerobics class two to three days a week. She springs up a flight of stairs like a fifteen-year-old.

Though also seventy, Jane is a couch potato. Turning on TV and moving her eye muscles are major exercises. Climbing a flight of stairs is like ascending Mount Olympus; a smoker, Jane pants and gasps for breath the whole way.

When we refer to "age," we ordinarily mean "chronological age." As you can see from Alice and Jane, however, "biological" age may not be related to actual age. If we take care of our bodies, we'll have a much better chance of completing our trip on the tightrope of life. If we become couch potatoes, smoke, drink to excess, and suck in toxins, as Jane does, we, too, will be old before our time.

Generally, biological aging means we can't respond to changes in our environment as we used to and our activity level declines. The rate of decline varies with each individual, as you can see from Alice and Jane. Who do you think will live the longest?

Regardless of your activity level, you're growing older daily! Those tell-tale wrinkles, that flabby skin, white hair, and aching joints are all definite signs that "we ain't what we used to be!"

Aging is a process that is inevitable, yet something we all dread.

As we careen merrily through life, with its ups and downs, curves and dips, we may ask ourselves, "Can we slow the car carrying us to the last stop of old age and, finally, death?"

Science believes so. Excited? Then read on!

Through our lives, we all secretly hope that someone will discover the true fountain of youth. Then we can all breathe a sigh of relief and rest assured that the aging process will stop and we'll all live forever.

Our life span should be at least 110 to 120 years, according to Professor Hans Kugler, Ph.D., in his book, *The Disease of Aging.*[1]

Factors that increase our life span not only add years to life, but also delay aging. According to Drs. Wilhelm Glenk and Sven Neu,[2] the most important life-extending factors are: 1) maintaining normal individual weight, 2) good nutrition (i.e., healthy diet plus supplements), 3) exercising, and 4) avoidance of pollutants and toxins. We would add a positive mental attitude to that list completing the Five-Step Jump-Start Plus Enzyme Program.

Laboratory studies show that several organisms (used in cell cultures) can multiply themselves forever. It is also possible to keep certain tissues and organs (such as a chicken's heart) alive, over an extended period of time, by using cell suspensions.

Every cell, however, will normally stop metabolizing and die within a certain time frame. Scientists now believe that cell lifespan is controlled, at least in part, by telomeres at the ends of chromosomes. These telomeres progressively erode with each round of DNA replication. Death occurs when they erode to the point that they can no longer protect the chromosomes. Most cells are not capable of producing error-free copies after they have divided forty-five to fifty times, so they self-destruct by inserting a receptor into their membranes. These receptors are the landing places for specific antibodies (immune complexes) which trigger a reaction resulting in an enzymatic disintegration of the age-weakened cells.

This mechanism is extremely important to the process of cell molting (the process of constant cell renewal taking place throughout our lives). Out with the old and in with the new, vigorous, perfectly functioning cells.

Scientists ask: Does life end as a result of accumulated errors or do genetics trigger a prechosen signal?

The people who suffer from Hutchinson-Gilford syndrome (premature old age) or Werner syndrome (premature adult senility) provide the tragic evidence that the aging process is genetically predisposed. These diseases involve premature senility (within a few years) at a young age, and cause rapid death from typical diseases of aging.

Genetic Coding

The genetic codes hold the key to aging and death. How? To answer this question, scientists from Japan and the United States crossed hamster cancer cells with human connective tissue cells. This caused genetic code changes enabling the hybrid cells to divide endlessly. They found that the human aging gene is within the first of our twenty-three chromosomes.[2]

Theoretically, gene technology could be used to alter human chromosomes and delay aging. Experiments have proven an increase in life span for tested molds, worms, flies, and mice. Further, this longer life expectancy is passed from generation to generation. This process, however, has not yet been demonstrated in humans. Presently, our best bet for living a long life is to come from a long line of ancestors who lived to a ripe old age.

There is a second factor in the aging process which also has to be considered—this being the possibility for error in the overall body metabolism.

We have no control over certain errors (metabolic abnormalities resulting from genetically determined enzyme deficiencies, for example). We can, however, avoid factors which weaken the body and cause premature aging. If we desire to live a longer and healthier life, it is important to embrace this concept and integrate it into our everyday life.

Aging and Digestion

As noted earlier, when people reach "black balloon" time, that is, forty years of age and over, there is a gradual reduction in hydrochloric acid in the digestive system. The result is difficulty in enzymatic digestion of foods, particularly protein foods. As we course toward and through the "golden years," steaks and other heavy foods are more difficult to digest. We are losing our enzymatic juices.

There is considerable evidence to show that, as you age, there is a decline in immune competence (this is characterized by losses in T-lymphocyte and other functions). Though some of these changes resemble those caused by malnutrition, a number of studies on well-nourished older rats indicate that even properly fed animals display weak immune responses. So is a weaker immune system just a side effect of old age?

We know that protein-energy malnutrition (at any age) can alter the proportion of T cell types, depress T cell function, and impair delayed hypersensitivity reactions, as well as thymic factor activity. These changes are strongly associated with increased susceptibility to infectious diseases. It is clear that severe malnutrition leads to impaired immune function in some older people. Equally clear is that improving their diets can at least partially correct immune function.

Alzheimer's disease affects approximately four million Americans. Estimates are that fourteen million Americans will have Alzheimer's by the middle of the twenty-first century. And, unfortunately, the longer you live, the greater are the changes that you will suffer from this progressive, degenerative disease. According to the Alzheimer's Association (AA), one in ten persons over sixty-five has the disease, and nearly half of those over eighty-five have Alzheimer's. This is also a particularly expensive disease. AA estimates that we spend $100 billion a year on Alzheimer's.[3]

Whether nutritional factors can alter the risk for this condition is not known. High concentrations of aluminum have been found in the brains of deceased patients, suggesting a relationship between aluminum and Alzheimer's disease. The brains of those who died from other causes don't show such high concentrations of aluminum. Despite these observations, there is no evidence that the use of aluminum antacids, antiperspirants, or cookware increases the risk for Alzheimer's. The significance of the increased brain aluminum concentration is unknown, but it might pay to be on the safe side and reduce your aluminum exposure.

Free Radicals

Human-made pollution and our unnatural way of life cause disturbances in the body (the large-scale production of "free radicals," for example). Free radicals are molecules which react chaotically and contribute to many diseases including cancer and circulatory disorders. They are thought to be one of the causes of many premature aging processes.

Free radicals are the "rust buckets" of our body. They contribute to the undesirable crosslinking of protein fibers in the connective tissue. The most obvious damage is the reduction of connective tissue elasticity, resulting in formation of wrinkles, sagging skin, and loss of the fatty layer beneath the skin.

Isn't it ironic that the very oxygen which is essential to production of energy in our bodies (a good event) can also lose direction, forming these free radicals? They can be a prime cause in such deadly free-radical diseases as cancer. This is balance—homeostasis—gone awry. We begin to tip to one side or the other in that balancing act we call life.

Antioxidants (such as zinc, selenium, vitamins C, A, and E, plus the enzyme superoxide dismutase) are used to fight these free radicals.

Free radicals oxidize cells to the point where they practically rust. It is impossible for normal metabolism to occur in these cells. If the destruction due to free radicals is not stopped, the whole body can be weakened to the point of illness and premature death.

They play a huge role in crosslinkage formation. Crosslinkages are the unwanted links between protein chains. These crosslinks occur in connective and other tissues, causing a reduction of connective tissue elasticity. The simplest test for crosslinkage is to raise the skin on the back of your hand (between two fingers) and release it. If the skin becomes flat immediately, the tissue has not yet been damaged, or aged, by crosslinking. "Crows' feet" at the corners of the eyes and frown lines on the forehead are clearly visible signs of crosslinkage in the connective tissue.

Giant protein chains make up the connective tissue, arranged in spirals and/or parallel to each other. Much like the spirals and wires of a spring mattress or trampoline, it is these giant protein chains that provide elasticity, strength, and tension.

Protein chain movement depends on interconnection and restriction of one another. Think of the firmly anchored strings of a harp. Each can vibrate elastically and then return to its original tension after the vibration stops. If two harp strings were connected, however, they would not be able to vibrate as freely, nor would they retain their elasticity.

The more injuries and other physiological problems our bodies sustain through the years, the more easily crosslinkages are produced, and the less easily our body is able to combat these crosslinkages (due to decreased enzymatic activity). The connective tissues, therefore, become stiff and rigid because the functional capacity of the involved tendons, muscles, nerve fibers, and blood vessels decreases.

We can avoid damage from free radicals and crosslinking by avoiding prolonged and unprotected time in the sun, ultraviolet (UV) and other radiation, nicotine (from cigarettes, cigars, etc.), saturated fats, industrial and exhaust gases, plus many other pollutants we inhale or consume. It is also better to eat fresh, raw vegetables and fruits daily (carrots, apples, and so on). We should only consume fats in the form of unsaturated fish or vegetable oils.

We can reduce the disastrous effects of free radicals by using "radical scavengers," or "antirust agents." These are antioxidants such as vitamins A, E, and C, zinc, selenium, enzymatic radical scavengers (such as SOD [superoxide dismutase], and hydrolytic enzyme combinations, including bromelain, papain, pancreatin, trypsin, chymotrypsin, amylase, and lipase).

Enzymes to the Rescue

When taken over a long period of time, certain enzyme mixtures break down and stop crosslinked protein chain formation. They contribute to tissue elasticity maintenance, and therefore, increase the capacity of body function.

As we age, our bodily enzymes decrease in amount and activity level. This decrease contributes to the symptoms of aging. For example, the phenomenon of greying hair is caused by a lack of the enzyme *tyrosinase*.

Plasminogen (a proenzyme) and plasmin (an enzyme) are crucial in sustaining the balance between blood clot formation and the breakdown of those blood clots. Plasminogen synthesis and synthesis of its activators is reduced with age, according to Dr. T. Astrup.[4]

This results in deposits of fibrin forming in the blood vessels. Increased fibrin formation seems to be associated with cholesterol and fatty material formation, which interfere with circulation. In addition to other circulatory problems, this can result in decreased brain, eye, ear, and kidney function, as well as heart attacks.

Pioneers in the use of enzymes, Drs. Max Wolf and Karl Ransberger treated the elderly therapeutically with proteolytic enzyme mixtures and vitamin E.[5] Many thousands of patients have profited from this therapy. Positive benefits included normalization of cholesterol levels and other serum lipids levels, plus an increase in blood clot-lysing activity.

Enzyme supplements, however, should be used as *supplements* and as only one part of the Five-Step Jump-Start Plus Enzyme Program. Don't forget the importance of diet, exercise, detoxification, and stress reduction.

Don't Make Mistakes

Gerontologists believe there are two ways to lengthen life span: changing the genetic code, and reducing metabolic errors. Organisms cannot operate completely free of error. There are approximately a billion cells in your body dividing, dying, and renewing at this very moment. Each cell transformation requires millions of steps which can only happen when certain enzymes are present. Because of these precise logistics, errors often occur.

The immune system's job is to fix mistakes. When we're young and healthy, our immune system is far more reliable. As years go by, this system, which is intended to repair body tissue, and maintain health, begins to make errors of sometimes serious proportion.

In technologically advanced countries, cancer, circulatory, and cardiovascular diseases are the primary causes of 80 percent of the deaths.

Circulation problems in the elderly result from decreased plasmin activation. This decrease prevents the right amount of blood liquefaction and the formation of fibrin gains the upper hand. This all happens in a sequence which is described in the cardiovascular section of this book. Enzymes' role in maintaining blood fluidity and reducing the formation of blood clots is essential to aging.

During the aging process, one of the most serious consequences is the possible formation of tumors. Cancer, for example, could actually be considered normal in old age, as could atherosclerosis (hardening of the arteries). All in the civilized world could acquire cancer if we survived long enough.

Cancer is a result of the aging process and reflects the body's inability to function in our current environment. Weaknesses in our immune systems caused by old age explain our growing susceptibility to infectious diseases.

When immune complexes remain in our system, the result is often the production of a new antibody, directed against the antibody in the immune complexes. It attaches itself to the complexes, attracting killer enzymes and tries to eliminate the undesired immune complexes from the body.

This second antibody (directed against the first) has to carry the correct identification to be able to attach itself (as the first antibody is specific for this antigen). Therefore, it imitates the foreign antigen. This is called "antigen mimicry." An anti-antibody of this type is also known as an "idiotype" (Greek, meaning *own type*).

Since this idiotype appears to look exactly like the antigen on the outside, the antibodies that work specifically on these antigens attack the idiotype. These are called "anti-anti-antibodies" or anti-idiotypes, which also form immune complexes with the idiotypes. If these are not kept in check, antibodies can form and be directed at the anti-idiotypes, which may happen in a weakened immune system.

If the chain reaction continues, it throws off the entire immune system, which, ultimately becomes paralyzed. This results in unchecked, uncontrolled activation in the wrong places. Fibrin is produced. Disorders initiated by the immune complexes occur and are identical to typical old-age disorders (premature chronic degenerative aging, for example).

The most effective way to slow age-dependent processes is by destruction and elimination of these immune complexes by macrophages and enzymatic action.

Hypothetically, each individual can control his or her rate of aging. It is the responsibility of each person to live a healthy life and eat a natural diet, to allow for healthy bodily functioning. In this way, we can reach our maximum individual life span.

Weight Loss

Did you know that the average American male consumes nearly 96 grams of fat every day? That equates to 864 calories from fat alone! And what's worse, about 34 of those fat grams are from saturated fat, implicated in heart disease and obesity. No wonder 35 percent of adult Americans are overweight! Being overweight is associated with elevated serum cholesterol levels, elevated blood pressure, and noninsulin-dependent diabetes, as well as being a risk factor for coronary heart disease.[6] Being overweight also increases the risk for gallbladder disease and some types of cancer, and has been implicated in the development of osteoarthritis.[6]

Insulin resistance is a well-known phenomenon among overweight individuals, and affects insulin's ability to act upon specific tissues. For reasons not completely understood, tissues of obese individuals often become insensitive to insulin. This is particularly true of muscle, adipose, liver, and brain tissue.

Insulin resistance is common among the obese and may be the reason for fat storage in this group of individuals.

Research tends to indicate that if we can improve the body's handling of insulin, we can help diabetes and weight loss.

Insulin resistance is well known in diabetes mellitus, lupus, obesity, polycystic ovary syndrome and other autoimmune diseases. A study of risk factors for coronary artery disease concluded that healthy persons with hyperinsulinemia and normal glucose tolerance are more prone to develop coronary artery disease, as compared with healthy subjects with normal insulin levels.[7]

But, insulin resistance can also affect people who are superficially healthy and normal, but have underlying pathological factors at work, according to Dr. Gerald M. Reaven.

Dr. Reaven compared plasma insulin levels with blood lipid levels and blood pressure in two groups of 247 healthy, non-diabetic, non-obese subjects with normal glucose tolerance levels. He found that "healthy" persons with elevated insulin levels had: 1) decreased HDL cholesterol levels, 2) increased LDL cholesterol levels, and, 3) high blood pressure as compared to "healthy" individuals with normal insulin levels.[8]

High blood insulin helps explain why diabetics (Type 2) can produce sufficient insulin (sometimes too much) but not use it properly and are two to four times more prone to develop heart disease than nondiabetics.

How do you become insulin resistant? One explanation for insulin resistance is a deficiency of biologically active chromium caused by: 1) inadequate dietary supply of chromium, 2) excessive chromium losses, or 3) the body's inability to convert this mineral into the biologically active Glucose Tolerance Factor (GTF) form. Chromium stimulates the activity of the enzymes that are involved in glucose metabolism.

In its biologically active form, GTF chromium acts in the body primarily as an insulin cofactor. Researchers believe that, by binding insulin to cell membrane receptor sites, GTF chromium increases insulin's ability to transport amino acids and blood sugar (glucose) inside cells for tissue synthesis and energy production.

The body tends to compensate by producing more insulin when insulin cannot work efficiently. The result is hyperinsulinemia (or high blood insulin) which can increase cardiovascular disease risk factors such as elevated LDL cholesterol, decreased HDL cholesterol, hypertension, and increased fat storage.

Although not all cases of insulin resistance are chromium-deficient, it appears that individuals who are insulin-resistant due to a chromium deficiency would benefit from chromium supplementation.

Eat Less and Live Longer

One step to living longer is by eating less. The Germans, for example, experienced a very limited diet in the period during and after World War II. Despite eating less meat and fat, they lived healthier and longer lives.[2]

The fact that human life expectancy has increased in the past few decades is not necessarily due to changes in dietary behavior, but probably to a decrease in infant mortality rates.

The relationship between food and life expectancy was examined by Dr. Roy Walford (a research immunologist at UCLA), who successfully increased the expected life span of mice two- to threefold by changing their dietary habits.[9] Dr. Walford had seen the damaging effects a rich diet had on the immune system. He is so convinced of the relationship between diet and life span that, to lengthen his own life span, he has reduced his caloric intake by almost one-half. The National Toxicology Laboratory in Little Rock, Arkansas has further demonstrated that thousands of rats and mice live two times as long when their caloric intake is reduced by 40 percent. All signs indicate humans could benefit from a similar reduction.

Numerous studies indicate that restricting our caloric intake can enhance our health and may extend our lifespan, as well.[10] One study on monkeys showed that reducing caloric intake by 30 percent over a ten-year period resulted in a twenty-point decrease in triglyceride levels and an increase in HDL (the good cholesterol).[11] High triglyceride levels, of course, are implicated in cholesterol buildup and heart disease.

Most Americans eat far too much. We eat the wrong food, at the wrong time of day, too quickly, and it is often too hot or too cold. We make all the possible mistakes which adversely affect normal digestion. We don't exercise enough, we damage our health by smoking, using other poisons, and exposing ourselves to pollutants.

It is estimated that by adjusting our lifestyle and reducing pollution, we could lower the amount of chronic diseases by two-thirds.

So, why are so many Americans overweight? The question may seem simple, but the answer is quite complex. Overweight reflects inherited, environmental, cultural, and socioeconomic conditions. Our weight tends to increase as we age until about age fifty for men and age seventy for women; then it declines. Overweight is particularly prevalent in minority populations, especially among women. Poverty is related to women being overweight, and the reasons for this are complex. Former Surgeon General C. Everett Koop conducted a survey to try and determine why those living on low incomes had such a high incidence of obesity.[12] He found that child-care responsibilities, neighborhood safety issues, and lack of sidewalks and recreational areas kept them from being physically active. In addition, their income levels made it difficult for them to purchase costlier and healthier foods, such as fresh fruits and vegetables.

Even though Americans now seem to be more health conscious—exercising more, and supposedly, watching what they eat, the prevalence of overweight has not declined among adults for two decades!

Energy Needs

As we get older, our energy needs decrease, probably because our physical activity decreases. But which comes first, the chicken or the egg? Further, our metabolic rate slows, thus diminishing our lean body mass.

Older adults need less food energy to maintain normal weight. Beginning at age fifty-one, slightly lower energy intake for adults is recommended, according to the Committee on Dietary Allowances.

Energy needs decline an estimated 5 percent per decade. The extent of overweight in adults over fifty-five years old also falls since energy intakes decline in parallel with needs. In spite of this, many older adults are overweight. This indicates that food intake has not decreased enough to compensate for their reduced expenditure of energy.

What can you do to reverse this trend? Exercise more, eat right, and put more enzyme-rich foods in your diet. You'll have more energy to do everything you need to do during the day and you'll improve your digestion, which is at the heart of many diseases.

Was it the famous French gastronome, Brillat-Savarin who said, "You are what you eat"? Do you really want to be a hamburger and fries?

The government says we should increase our intake of complex carbohydrates and fiber-containing foods to five or more daily servings for vegetables and fruits, and six or more daily servings of grain products.[6] Vegetables (including legumes, such as beans and peas), fruits, and grains are good sources of complex carbohydrates and dietary fiber, as well as several enzymes, vitamins, and minerals. These foods are also generally low in fat and can be substitutes for foods high in fat. Dietary patterns with higher intakes of fresh vegetables (including legumes), fruits, and grains are associated with a variety of health benefits, including a decreased risk for some types of cancer.[6]

In fact, those populations which consume diets rich in fresh vegetables, fruits, and grains have significantly lower rates of cancer of the colon, breast, lung, oral cavity, larynx, esophagus, stomach, bladder, uterine cervix and pancreas.[6]

Is your belt getting a little tight in the waist? Holding your breath to button that top button? Afraid to hop on the scale? Then you're not alone. Millions of Americans just like you are attempting to fight the "battle of the bulge." Cycling, stair climbing, hiking, running, aerobics, and pumping iron have become the norm in today's society.

An increased knowledge of diet and nutrition coupled with the "thin is in" mentality has created a society bent on perfection. This phenomenon has opened the doors to a fiercely competitive $36 billion weight loss, diet, and exercise industry.[13] Americans buy into this market by foolishly signing up for diet plans that have no scientific basis in the hope of shedding those extra pounds. Television talk-show host, Oprah Winfrey, for example, has known the thrill of weight loss (while on a liquid diet) and the agony of watching those pounds creep back on when she returned to normal eating. Only when she followed a nutritious diet and exercise plan did she finally achieve permanent success.

But what causes those unsightly bumps, sags, and rolls to appear in the first place? Why is it that some people live their lives trying to gain weight and some simply cannot lose an ounce? The answer to these questions may lie in our little friend, the enzyme.

We know that enzymes are the catalysts to all reactions. They begin all metabolic reactions and are therefore necessary to begin chain reactions from the mouth to the brain. Within the brain lies our appetite control center, the hypothalamus. If our hypothalamus is working correctly, we will be able to maintain a proper weight, or lose weight when we diet. It sends a message through our nervous system when to eat and when to stop eating.

Disturbances in lipid and carbohydrate metabolism in obesity have been well-established. Functional adaptation of the pancreas to diet has been well-demonstrated in rats and dogs. A high-protein diet increases the pancreatic content of protease and a high-carbohydrate diet increases pancreatic amylase.

One study, conducted by Japanese researchers, studied the serum levels of pancreatic enzymes in lean, as compared to obese, subjects.[14]

The researchers found that serum amylase and trypsin, but not lipase, were significantly higher in lean subjects than in obese subjects. Low serum amylase was associated with low protein, fat, and carbohydrate intake (per kilogram of body weight), while low serum trypsin was associated with low carbohydrate intake (per kilogram of body weight). When the obese subjects lost weight, their serum amylase levels increased.

In another study, researchers assembled two groups of rats. They cut the nerves on one side of the hypothalamus in the rats in Group A. The rats starved to death, even though food was right in front of them, because they didn't get the message to eat. The researchers then cut the nerves on the other side of the hypothalamus of the rats in Group B. These rats gorged themselves to death because they didn't get the message to stop eating.[15]

Why is it that some people don't get the message to stop eating and become obese, and some don't get a message to eat at all and become anorexic? In many situations, nerve cells become damaged through lack of nutrition and healthy cells have a hard time rejuvenating.

But how does the hypothalamus get the message to be hungry or satisfied?

Chemical messengers, called neurotransmitters, which are synthesized in the brain from our nutrient intake, carry signals to the hypothalamus.

When everything works correctly, and our neurotransmitters receive adequate nutrients, our weight stays constant. But, what happens in those people who either become obese or extremely thin and cannot gain or lose as needed? The neurotransmitters were not given adequate nutrients.

The brain cannot provide nutrients to neurotransmitters by itself. Our nutrient source is through the blood in the brain's capillaries. Consequently, there is no way for the brain cells to adequately feed neurotransmitters unless we consume a healthy diet.

Neurotransmitters

In order to properly supply our neurotransmitters with nutrients, six factors must be present: enzymes, coenzymes, vitamins, minerals, glucose, and amino acids. Enzymes are the catalysts to all reactions in the body, and most importantly in weight control, to those in the brain. Enzymes are made of amino acids which are breakdown products of the protein we eat. These amino acids are precursors to neurotransmitters and are found in the blood *if we assimilate our protein properly.*

Some precursors, or amino acids, which may be familiar to you are tryptophan and tyrosine. Tryptophan and tyrosine work very closely with a neurotransmitter called serotonin.

When tryptophan enters the brain at an accelerated rate (say, after protein consumption), serotonin is synthesized. Research has shown that serotonin can suppress carbohydrate consumption in rats and obese human beings.[16,17] Conversely, low serotonin levels encourage a person to eat more carbohydrates and less protein. If a faulty equalizing system such as this is present, the person might constantly crave carbohydrates.

Obesity is one of the biggest diet-related problems in the United States. It affects 35 percent of all adult Americans and 12 percent of all adolescents ages twelve to seventeen.[18] Historians have looked for clues as to the evolution of obesity and concluded that it did not appear with its present frequency until the eighteenth and nineteenth centuries in upper-class England.[19] It was here that the nobility showed their wealth and power by gorging themselves with food the peasant could not afford. Being overweight and obese were actually considered to be signs of health in this country until the twentieth century. Some societies actually still believe that, to be healthy, one must be a little on the "plump" side.

Causes of obesity are not really known yet, but some contributing factors may be 1) heredity, 2) overeating, 3) altered metabolism of fat tissue, and 4) decreased physical activity without lowered food intake. Poor health is not necessarily a consequence of obesity, but excess body fat poses risks to many Americans. The exact contribution of obesity to illness and premature death is still not known, but obesity undoubtedly contributes to premature mortality, especially when associated with blood cholesterol levels, high blood pressure, or diabetes.

Unfortunately, no clear cure for obesity has been discovered, but it is helpful to know that we have discovered some causes. In an enzyme deficiency, for example, some of the most common symptoms are fatigue, premature aging, and weight gain. Research strongly suggests that if we were to adjust our eating habits to consume the optimum, all-around diet, accenting proteins, vitamins, minerals, and enzymes in their correct biological proportions, our life spans and vitality would skyrocket to unimaginable new proportions.

It is therefore important to understand the food we eat and its enzyme content. This is because consumption of complex carbohydrates in meals and snacks has been recommended for weight loss.

Complex carbohydrates maintain glycogen stores in our bodies for optimum energy levels. Also, eating carbohydrates triggers a significant increase in brain serotonin, followed by a decrease in hunger impulse.

It is important, however, to remember that we are all different. Each individual's metabolism is based not only on his or her activity level but also on genetic background. What works for you might not work for me, and vice versa. For some people, consuming a diet high in complex carbohydrates and low in fat helps them lose weight. Others can only lose weight if they eat fewer carbohydrates and increase their intake of protein and fat. A number of current diets seem to support this theory.

But regardless of the method you choose to lose weight, do it naturally and avoid drugs as much as possible. One of the latest "cures" for obesity is lipase inhibitors. These drugs—which go by a variety of names, including Orlistat®, tetrahydrolipstatin, and Xenical®—actually inhibit the activity of the enzyme lipase. By doing so, they reduce the digestion and absorption of the fat in your diet. Unfortunately, your body needs a certain amount of fat in order to absorb the fat-soluble vitamins A, D, E, and K. In addition, these drugs come with side effects including diarrhea, flatulence, and other gastrointestinal problems. Wouldn't it be easier and safer to just cut down on the amount of fat that you eat?

Another strategy to decrease overeating is by eating low-calorie foods that are high in fiber, thus providing bulk and the feeling of fullness. There are hundreds of food sources that fit this description, including corn, potatoes, yams, beans, pumpkin, and squash. As fresh fruit and vegetable consumption increases, enzymatic activity within the body will also increase, catalyzing all bodily functions to new heights of vitality.

EXERCISE FOR THE HEALTH OF IT

As with any healthy diet, exercise is of the utmost importance. The human body evolved in eras when great energy was expended in obtaining food, in protection, and in surviving the elements. In modern, affluent societies, it is usually necessary to build a program of energy expenditure to assure optimal function for bodies that historically developed processes for such energy exchange. An exercised body is more likely to remain healthy than one which remains sedentary.

Regular exercise: 1) helps maintain optimal body composition; 2) improves the possibility of losing weight when necessary; 3) increases the efficiency of muscle fibers to produce energy; 4) increases efficiency of hormones (insulin, lipoprotein lipase, epinephrine) to regulate energy metabolism; 5) decreases the production of lactic acid, which interferes with energy production; 6) strengthens the heart, lungs, and circulatory system; 7) increases levels of HDL over LDL cholesterol and decreases serum triglycerides; 8) raises the rate of basal metabolism; and 9) helps control appetite.

All things considered, a well-balanced life, including a high-carbohydrate, high-fiber diet and consistent exercise, will boost human vitality to new heights.

Numerous studies have examined the effects of excessive body weight on mortality. In most studies, mortality increased with increasing weight. Relative body weights 20 percent over the desirable norm are associated with diabetes, digestive diseases, and cardiovascular disease. The higher the relative weight, the greater the risk for these conditions. Many other serious conditions such as gallstones, sleeplessness, osteoarthritis (and other disabling disorders of locomotion), bear a direct relationship to obesity. With the immense list of risks involved in obesity, it is quite comforting to know that we can take charge of our lives through the use of enzymes. With regard to the increasing weight obsession, Americans can rest assured that enzymes provide a crucial key to unlock the door to health and well-being.

Living Longer

Science has proven that we can slow the aging process, look better longer, and increase our life span if we follow certain rules. These are:

1. Use enzymes and antioxidants (such as vitamins A, E, and C, plus zinc and selenium).

2. Control our diet.
3. Increase exercise.
4. Maintain normal body weight.
5. Avoid toxins and other pollutants.

By following these rules, we can slow down the roller coaster. All aboard! Get your ticket to a happier, healthier, more productive life.

References

1. Hans Kugler, *The Disease of Aging*, Los Angeles: Keats Publishing, Inc., 1983.
2. Wilhelm Glenk and Sven Neu, *Enzyme Die Bausteine des Lebens Wie Sie Wirken, Helfen und Heilen*, Munich, Germany: Wilhelm Heyne Verlag, 1990.
3. Alzheimer's Association. Retrieved from http://www.alz.org/facts/rtstats.htm.
4. T. Astrup, *Biology of Plasmin*, Stuttgart, Germany: Schattauer, 1970.
5. Max Wolf and Karl Ransberger, *Enzyme Therapy*, Los Angeles: Regent House, 1972.
6. *Healthy People 2000.* Washington, D.C.: U.S. Department of Health and Human Services, Public Health Service, 1991, 112–118.
7. Zavaroni, I., et al., "Risk Factors for Coronary Artery Disease in Healthy Persons with Hyperinsulinemia and Normal Glucose Tolerance," *New England Journal of Medicine* 320: 702-706 (1989).
8. G. M. Reaven, "Banting Lecture 1988: Role of Insulin Resistance in Human Disease," *Diabetes* 37: 1595–1607 (1988).
9. Roy Walford, *Retardation of Aging and Disease by Dietary Restriction*, Springfield, Il.: Charles C. Thomas, 1988.
10. R. B. Verdery and R. L. Walford, "Changes in Plasma Lipids and Lipoproteins in Humans During a 2-Year Period of Dietary Restriction in Biosphere 2," *Archives of Internal Medicine* 158(8):900–906 (27 April 1998).
11. R. B. Verdery, D. K. Ingram, G. S. Roth, and M. A. Lane, "Caloric Restriction Increases HDL2 Levels in Rhesus Monkeys (*Macaca mulatta*)," *American Journal of Physiology* 273(4pt 1):E714–719 (October 1997).
12. *Shape Up America!* Press release, Retrieved from http://www.shapeup.org/surveys/barrier.htm.
13. Ruth Papazian, "Never Say Diet," *FDA Consumer* 25:8 (Oct., 1991).

14. Takaharu Kondo, Tetsuo Hayakawa, et al., "Serum Levels of Pancreatic Enzymes in Lean and Obese Subjects," *International Journal of Pancreatology* 3:241–248 (1988).
15. Ruth Yale Long, *Home Study Course in Nutrition*, Los Angeles: Keats Publishing, Inc., 1989, 12–15.
16. N. Orthen-Gambill and R. B. Kanarek, "Differential Effects of Amphetamine and Fenfluramine on Dietary Self-Selection in Rats," *Pharmacology, Biochemistry and Behavior* 16:303–309 (1982).
17. J. Wurtman, R. Wurtman, et al., "D-Fenfluramine Selectively Suppresses Carbohydrate Snacking by Obese Subjects," *International Journal of Eating Disorders* 4:89–99 (1985).
18. CDC Fastats. Retrieved from http://www.cdc.gov/nchswww/fastats/overwt.htm.
19. *Statistical Abstract of the United States*, Washington, D.C.: U.S. Department of Commerce, 1992, 271.

CHAPTER 10

Beauty Is Only Skin Deep

HAVE YOU looked into the mirror lately only to gasp and painfully acknowledge yet another wrinkle on your face? Do you find yourself describing the wrinkles around your eyes and mouth as "crows' feet" or "laugh lines"? Join the club. We're all getting older, day by day, and it seems as though our bodies and skin have joined in the foot-race against time. We fumble through lotions, creams, and gels on store shelves, hoping our increasing age lines will "disappear overnight" or "significantly reduce" in five to seven days. Is there any truth to cosmetic claims or are we destined to grow old, wrinkled, and saggy?

Your skin, which has covered you since the day you were born, is not the same skin that covered you seven years ago, or even yesterday. It's constantly regenerating. The fat beneath your skin is not the same fat that was there a year ago, nor will it be the same a year from now. Your body is constantly changing and your skin is going right along with it.

So, what do we do? We spend incredible amounts of money and time trying to make ourselves look younger and more vital. Plastic and other body-contour surgeries have become the norm nowadays. In fact, according to the American Society of Plastic and Reconstructive Surgeons, the number of cosmetic surgeries performed in the United States more than doubled between 1992 and 1998.[1] From facial reconstruction to liposuction, the quest continues for never-ending, eternal youth and beauty. But, perhaps in our superficial search, we have overlooked one minute, yet obvious, key to the puzzle: enzymes.

Our skin, the largest and heaviest of the body's organs, weighs about six pounds (on the average). Skin is composed of the epidermis, the outer layer, and the dermis underneath that. A fatty subcutaneous layer helps to absorb shocks and to insulate the body. The epidermis is constantly renewing itself, with a turnover rate of about three to four weeks. Just above the dermis, millions of epidermal cells are formed daily. In turn, they are pushed upwards toward the outside by newly formed cells. The epidermal cells change from round cells having a jellylike consistency to flatter and harder cells. As the epidermal cells migrate upward, near the harsh environment of the outside world, they die and slough off.

The skin becomes thinner and more transparent as we age. It sags, and the subcutaneous layer loses fat. Unfortunately, the resilience of the elastic skin fibers is lost as well.

The primary component of the dermis is collagen, one of the body's most widespread proteins. Skin gets its elasticity, smoothness, and strength from a fine net of elastin (an essential component of connective tissue) with an interlacing of collagen fibers.

Because there is no blood supply to the epidermis, cell nourishment depends on diffusion from the dermis. Damaged elastin and collagen are reactivated by enzyme combinations in fibrin circulation, also helping to protect them against further attack by degrading enzymes (collagenase and elastase) and by free radicals.

These enzymes bind to collagen fibers, realigning them in a more youthful, undamaged form, and act as free-radical fighters.

Table 10-1. Ten Reasons Why Skin Ages More Quickly

1. Climate
2. Drugs
3. Environmental pollution
4. Genetics
5. Nutrition
6. Poor circulation
7. Radiation, sun
8. Smoking
9. Stress
10. Weight swings

Enzymes and enzymatic activity are intimately related to the vitality and health of our skin. Decreased enzyme activity and numbers give rise to the definite signs of an aging skin (blemishes, discoloration, premature aging, sagging skin, wrinkles, and lifelessness).

Your skin is a direct reflection of your health. It reflects such factors as genetics, diet, nutrients digested and absorbed, elimination, hormones, your skin-care program, your emotions, and your diet. Although we can't influence our genetics, we can make changes in our diet. But that's not enough. According to Corey Resnick, N.D., "Even the best diet in the world won't make your skin healthy and beautiful if your body doesn't utilize nutrients from your food."[2] As we now know, enzymes in fresh foods and in supplements assist our body in better digesting and absorbing our food, as well as in eliminating the waste.

Further, when we experience an enzyme deficiency, key nutrients remain locked in our foods. Digestive enzymes unlock the nutrients and make them available to our bloodstream and our skin.

As discussed earlier, our body produces and requires protein-, fat-, and carbohydrate-digesting enzymes, in order to obtain fats, proteins, and carbohydrates from food to form building blocks for our skin.

For example, specialized proteins maintain the integrity of our supporting tissue and furnish normal elasticity for our skin. The interlocking of elastin with collagen strengthens the elasticity of the skin and helps it to remain smooth. Enzyme deficiencies can cause loss of muscle tone, sagging skin, and loss of elasticity.

Enzymes fight aging skin not only by eliminating free radicals and preventing crosslinking but also by improving the blood supply to the skin's dermal layer. This allows more nutrients to reach the tissue. Circulation slows down during the aging process; enzymes can help maintain better circulation.

Further, enzymes are beneficial in fighting food allergies, intestinal toxemia, and immune-deficiency problems that can be reflected in those mirrors called your face and skin.

Enzymes can also be used topically to improve the health and appearance of your skin. A number of products containing enzymes, including creams, lotions, moisturizers, exfoliants, face-lift formulations, and skin dips, are available. The most popular enzymes seem to be papain, bromelain, and superoxide dismutase. These and other enzymes are effective at treating numerous skin conditions, including acne and wrinkles, when used topically.

Conclusion

In order to slow down the skin-aging process, we must take control of our lives. Enzymes—taken internally and applied topically—can help you achieve younger-looking skin. Check the following list for a possible tip-off to skin problems:

1. Enzyme deficiencies
2. Food allergies
3. Free-radical overactivity
4. Intestinal toxemia
5. Nutritional deficiencies
6. Toxin intake (e.g., coffee, alcohol, smoking, etc.)

References

1. ASPRS National Clearinghouse of Plastic Surgery Statistics press release, April 28, 1999.
2. Roberta Wilson, "Plant Enzymes Can Make Your Skin Healthier," *Let's Live* (April 1993): 61–64.

CHAPTER 11

Bounce Back Faster from Injuries

OK, so you feel good, you look good. You're on the Five-Step Jump-Start Plus Enzyme Program.

But, yesterday, while cutting roses, you stuck a thorn in your finger, the thorn broke off, and you can't get it out. The area is swollen, it hurts, is hot to the touch, and is now turning red. You've got an inflamed finger. But, what's happening? What is inflammation?

Inflammations are unpleasant, often painful, and range from whiplash injury to back pain, from acute appendicitis to a sore throat, from renal inflammation to that thorn in your finger. We use words ending in "itis" to describe almost every disorder involving inflammation (i.e., arthritis, pancreatitis, dermatitis). These conditions respond well to supplemental enzyme treatment.

Anything that injures body tissue can cause inflammation. The list is endless and includes burns, wounds, X rays, viruses, and bacteria, to name a few.

Each step of the inflammatory process is necessary before healing can be completed. The first step is the *reaction* to injury (as well as the fight against any pathogen which could lead to infection). The second step is *repair* of the damaged tissue, and the third is *renovation* of the damaged area.

These three phases (reaction, repair, and regeneration) are present in every inflammation and sometimes gradually overlap.

Reaction Phase

Immediately after that thorn broke off in your finger, your capillaries (very small blood vessels) contracted for about half a second (this can be longer, depending on the stimulus). After contraction, they began to dilate. This stage can last up to seventy-two hours.

The area around the thorn began to swell, as fluid containing healing cells poured into the area (this also gave the tissue its characteristic redness and heat).

The body's job is to get macrophages, platelets and lymphocytes (your white blood cells) to the wound to clean it up and begin the repair process. It does this by activating messengers in the capillary blood vessels near the injury, who, in turn, alert pain receptors. The nerves tell the brain, "Hey, I'm hurt." This lets the brain know it should do something about the injury.

These messengers enlarge the capillaries, allowing more fluid to enter the wounded area.

The blood vessels surrounding the damaged or destroyed tissue become more permeable. The helper cells (which destroy or remove any enemy) then seal off the damaged region, break up any tissue debris, and clear up the inflamed area. The helpers circulate in the blood fluid of the injured area and pour through the spaces between the cells of the blood vessels.

Fluid which sometimes weeps from a wound may contain enemies (bacteria or viruses). Bacteria, for instance, can attack this fluid and cause infection—you know this as pus.

Because the damaged area has been sealed off, blood flow is stopped. Fibrin clots are formed, obstructing the blood flow, increasing exudation and swelling. This isolating barrier is called the *inflammatory membrane*. It is at this stage of inflammation that the therapeutic use of anti-inflammatory enzymes is the most effective.

The ability of certain enzymes to "chew up" protein (fibrin) means that enzymes can dissolve these clots, allowing oxygen to reach and revive the tissue cells. The result is that the waste products floating in the fluid are excreted (via the lymphatics). When cells are injured, they release proteolytic enzymes which break proteins down into harmless components. These, in turn, increase the permeability of the capillaries.

REPAIR PHASE

Now that your body has reacted to the pain of the rose thorn, it begins the reparative process. This can take seventy-two hours to six weeks, depending on the nature of your injury. Proper circulation in the area is restored which will help the body eliminate any toxic by-products of tissue decomposition. Improved circulation will also bring nourishing material and your body's natural enzymes, such as plasmin, to the area.

Capillaries bring oxygen and nutrients into the damaged area, which leads to the production and deposition of collagen. Collagen is a protein, the major component of your intervertebral disks (the pads between the vertebrae) and the majority of your connective tissue.

RENOVATION PHASE

The last phase can last from three weeks to twelve or more months (again, depending on the seriousness of your injury). During this phase, new cells are formed in an attempt to fill-in the injured area.

In the third phase, the collagen that was deposited in the second phase is improved, making it stronger and better able to withstand the stresses imposed upon it.

How Proteolytic Enzymes Work

The process of any inflammation is governed by numerous enzymes, especially the body's own proteolytic enzymes. They eliminate the inflammatory debris and initiate the restitution.

This process can be supported and accelerated by the use of supplemental proteolytic enzymes. They keep the pathological process from spreading and considerably reduce the duration of the disease. Proteolytic enzymes work thrombolytically to break up blood clots in two ways. First, they activate the body's own proteolytic system. Secondly, they have fibrinolytic activity of their own.

Proteolytic enzymes accelerate the inflammatory process necessary for the healing of the wound. This acceleration means, on the one hand, that the work of damage control, damage repair, and new tissue construction is carried out more forcefully and precisely, and thus, completed more swiftly. On the other hand, it also means that there can be a temporary increase in the visual and sensory effects produced by the inflammation (that is, more redness, swelling, heat, pain). This is not bad. It is a sign that the body is functioning; the rescue and repair teams are hard at work at an increased activity level.

Table 11-1. The Effects of Proteolytic Enzymes

1. Inhibit inflammation without immunosuppression.
2. Reduce pain and edema.
3. Inhibit thrombi formation and stimulate thrombolysis.
4. Improve the blood supply and therefore, the nutrition of the tissue.
5. Improve blood circulation.

With enzyme therapy, pain stops quickly and the duration of inflammation is rapidly dimished. In addition, wounds heal rapidly with reduced scar formation.

When your cells are injured, they die and become "foreign" to the body. These dead cells must be removed from the tissue before they can lead to infection. Enzymes in your body degrade and remove these foreign cells.

Let's revisit the blood vessels after an injury. What's happening?

No matter how your injury occurred, whether in the backyard, in the office, or while playing sports, a sprain is a sprain and a bruise is a bruise. . . . Like death and taxes, they're all the same process!

The flexible capillaries fan out, like a web, servicing every cell in the body. Think of these tubes (no matter how small) as constructed of bricks with cement holding them together. In an injury, the cement weakens, the bricks separate, and fluid escapes. The fluid from the capillaries forces its way into the surrounding tissue causing a traffic jam and congestion. There is more material (fluid, waste, etc.) than the area can handle. Therefore, we have increased pressure and swelling with the resulting heat and pain. Remember this concept, because this is what happens after an injury and can be the basis for all disorders involving inflammation.

So, now it is important that the body chews up and disposes of the garbage. (Where's Waste Management when you need it?) It is also critical that the repair crew seal up the cracked and broken brick tubes.

To do this, proteolytic enzymes can call on bioflavonoids (and possibly vitamin C) to help them with the sealing (cementing) process. Bioflavonoids render your capillary walls more impermeable, not permitting fluids to pass through their coating, and thus quickly reactivating the pumping action of capillary extremities. This is how certain bioflavonoids work to prevent and stop excessive bleeding from injuries (and bleeding gums, as well).

When an individual is first injured, a series of metabolic processes (inflammation) take place. One of the major concerns in inflammation is capillary permeability.

Capillaries (the smallest blood vessels) are responsible for carrying nutrients (including oxygen) to the cells and removing wastes (such as carbon dioxide). Injury damages some of our capillaries, making them temporarily incapable of carrying fluid to and from damaged tissues. The unwanted result is pain, swelling, redness, heat, and you out of action (the classic signs of inflammation).

Muscle cramps are caused by the same process, that is, a sudden lack of enough oxygen.

It is critical that something be done. The sooner your capillaries resume their pipeline activity, the sooner bruises, swellings, and pain disappear, and the sooner you get back into action.

Capillaries are the "fields of labor" where the essential activities of the circulatory system are carried out. The importance of a healthy heart, arteries, and veins cannot be minimized—they maintain an adequate rate of blood flow through the capillary beds.

An inflammation may be localized at the site of the injury or it may extend throughout the body in such tissues as the brain, nerves, heart, liver, lungs, spleen, adrenals, muscles, and the intestinal tract, on a generalized basis. Tissue repair, wound healing, and the increased progression of the inflammatory process are dependent upon accelerated cellular metabolism, additional supplies of vitamin C, bioflavonoids, zinc, enzymes, and other nutrients.

The addition of supplemental proteolytic enzymes, zinc, bioflavonoids, and vitamin C has many beneficial effects. Injury recovery rate is much more rapid, muscular cramps seem nonexistent, and swelling (caused by contusions or joint injuries) is minimal and subsides rapidly. These factors are very important in the care of injuries.

The Jump-Start Concept

Repeated minor (as well as major) injuries increase the requirements of essential nutrients during continued stress and physical exertion. This means your diet and supplements must compensate the body for nutrient losses and body adaptation.

Diet is important when injuries occur. Eat 50 to 75 percent raw foods (for increased enzymatic action). Drink fresh juices and herbal teas. Avoid junk foods, colas, refined white sugar, refined flour, in fact, all refined food products (which increase body stress). Follow the Five-Step Jump-Start Plus Enzyme Program.

Proteolytic enzymes have been successfully used in a wide variety of injuries with resulting rapid recovery.

Dr. Cirelli[1] used bromelain to treat patients (most of whom were police or fire fighters) for inflammation edema with accompanying contusions, abrasions, hematomas, ecchymoses, sprains, strains, lacerated and perforated wounds, operative wounds, fractures, cellulitis, hypostatic and diabetic ulcers. The study found 2.5 percent of the patients healed more rapidly than expected.

Bromelain was also used in a study involving fifty-nine patients with blunt injuries to their musculoskeletal systems.[2] The orthopedic surgeon who conducted the study measured the patients' progress on the days of their injuries and on five subsequent dates. He found that the patients taking bromelain had a clear reduction in swelling, in pain at rest and during movement, and in tenderness.

Dr. Fulgrave[3] communicated his experiences with sport injuries. He feels that, in many cases, the usual forced inactivity for a two-month rest period after major injury could be reduced to two weeks. Also, sprains and strains to knees or muscles can be impressively controlled by proteases. In some parts of Germany, prize fighters are directed to take enzymes prophylactically before the fight to prevent the results of severe injuries.

Dr. Worschhauser[4] reviewed conservative systemic enzyme therapy for sports injuries. At the time of his study in West Germany, sports injuries represented more than 10 percent of all accidents (approximately three million a year). Nearly 80 percent of these injuries were soft tissue lesions, such as contusions and compressions. These tissue lesions and vascular ruptures caused swelling and pain. Immediate enzyme therapy, including cooling, compression, and discontinuation of sports, was obligatory.

Some patients in his study received drugs, often with considerable adverse reactions, depending on the class of drugs. The patients receiving natural enzyme preparations, however, experienced a very low incidence of side effects. In addition, their recovery time was reduced, allowing them to participate in the sports activity sooner.

Studies on subjects competing in karate, as well as ice hockey players in the German Federal Division over two playing seasons, showed earlier resumption of the sports activities, dependent on the enzyme dosage. Because of these results, Worschhauser concludes that the prophylactic use of enzymes should be taken into consideration to prevent tendon pathology and muscular lesions.

Dr. Rathgeber[5] studied the use of proteolytic enzymes (chymoral) in treating certain sports injuries. A double-blind trial with either chymoral or a placebo was completed on forty-three athletes having sustained injuries due to accidental trauma in sport. According to Rathgeber, statistically significant results were achieved in three of the four parameters used (namely bruising, return to function, and fitness to resume play). The patients recovered rapidly when treated with chymoral which helped them to maintain their morale, personal fitness and skill levels.

Dr. Holt[6] used papain (*Carica papaya*) in a two-year study investigating the treatment of injured athletes. He found that bruises healed rapidly when proteolytic enzymes were used in anti-inflammatory therapy.

In another study, proteolytic enzymes were used prophylactically on sixty-four football players. Dr. Cichoke and Leo Marty analyzed the recovery rate from soft tissue athletic injuries during the 1980 spring football season.[7] This highly controlled, double-blind study (conducted through the Department of Sports Medicine, Portland State University) compared the rate of recovery in two groups. The first group received an orally ingested, proteolytic enzyme mixture.

The second (control) group took a placebo (sugar tablet). Neither the participants, nor the doctors dispensing the tablets, knew which group was receiving the enzymes. Only the manufacturer knew. Test results indicated that time lost from play was *reduced* in the enzyme group when compared to the placebo group.

Results of this study indicate that it is possible to obtain some protection from injury by taking enzymes prior to competition. In addition, should injury occur, enzymes help to reduce recovery time by approximately one-half to one-third.

Another contact sport, karate, is a martial art in which injuries occur in every area of the body, much like football. With this in mind, Dr. Zuschlag used the following double-blind investigation.

Ten karate fighters of both sexes were treated prophylactically (before fighting) with enzyme tablets. The second group of ten received a placebo. All 20 athletes had injuries comparable in severity. Findings from the karate study revealed certain highly significant results, as shown in Table 11-2.

Table 11-2. Results of Karate Study

	PARAMETER	DURATION
	Experimental Group	Control Group
Hematoma	6.62 days	15.59 days
Swelling	4.25 days	9.82 days
Restriction of movement	5.04 days	12.62 days
Inflammation	3.83 days	10.56 days
Unfit for training	4.18 days	10.23 days
Unfit for work	2.00 days	5.29 days

Source: J. M. Zuschlag, "Double-Blind Clinical Study Using Certain Proteolytic Enzyme Mixtures in Karate Fighters," working paper (1998).

The results were impressive in that hematomas developing in athletes treated with enzymes disappeared within 6.62 days. On the contrary, athletes in the placebo group took 15.59 days to recover.

The swellings suffered by the enzyme group subsided after 4.25 days, while those in the placebo group took 9.82 days. Restrictions of movement as a result of pain and injury disappeared after 5.04 days in the enzyme group, but lasted more than 12.62 days in the placebo group. When the injuries suffered during the karate competition became inflamed, these had subsided after only 3.83 days for the enzyme group, but took 10.56 days to do so in the placebo group.

Dr. Rahn[8] studied the effectiveness of an enzyme mixture in trauma, especially in knee surgery. Eighty patients participated in this double-blind study. The results showed that there was rapid subsidence of wound edema (swelling) in the enzyme group, which is a major prerequisite for early functional mobility therapy. On average, patients in the placebo group were able to bend the knee on the ninth day after surgery, while patients treated with enzymes were able to bend the knee joint to 90° on the seventh day after surgery. Therefore, the patients could stand and walk earlier. Patients in the enzyme group showed faster improvement in symptoms such as edema, pain, and limitations of mobility and flexibility from the first to the seventh day after surgery, as well.

Dr. Lichtmann[9] treated professional boxers and found that an injection of trypsin in the buttocks immediately after injury allowed black eyes and bruises to subside in one to three days, rather than the usual ten to fourteen days. Since trypsin has little effect once discoloration is seen, he stresses early treatment.

Several physicians have noticed excellent results in treating ischemia. Ischemia is the deficiency of blood either because of actual obstruction of a blood vessel or because of a functional constriction. A patient with bimonthly recurring ischemia participated actively in his sport without an attack for over a year while taking enzymes prophylactically. After interrupting the therapy for three weeks, the symptoms returned, but disappeared immediately after resumption of the enzymes.[10]

A world-class swimmer (a butterflyer) developed shoulder problems. The butterfly stroke involves repeatedly raising the arms out of the water, above the head and pulling down very hard. It puts a tremendous pressure on the shoulders. The swimmer experienced extreme pain in both shoulders and was very concerned because the national championships were within a month's time. He wanted to qualify for the national swim team. He was put on proteolytic enzymes. Within a week, the pain had almost gone, in spite of continued workouts. By the second week, he was totally pain free, was doing his former stress workouts, and, as a result, scored high in the national swim championships.

In another instance of enzymes' benefits, a high-school high-jumper was experiencing tight leg muscles and having difficulty jumping. One reason he could not increase his jump was that he had experienced great pain in taking off (doing the Fosberry flop—a high-jump technique). He began taking proteolytic enzymes on a daily basis. Within a week's time, he set a new high jump record, increasing his best jump four inches by jumping six feet, eight inches. Ultimately, he cleared seven feet, two inches.

Mike played middle linebacker for a local high school. His left knee was violently injured during a critical game and he was carried off the field. His knee immediately began to swell. The team physician, after examination, stated Mike was out for the season.

Immediately after the game, Mike began taking high doses of proteolytic enzymes. He continued the enzymes over the weekend, and on Monday, returned to the team physician for examination. The physician, having seen Mike on Saturday, could not believe the rapid improvement in his condition. The doctor was amazed that Mike was not only walking, but he was almost pain free. The doctor released Mike and, that Tuesday, he returned to practice and continued taking enzymes. By Friday (the next game), he felt no pain whatsoever in his knee and the swelling had totally disappeared.

Joe was a long-distance runner who developed knee problems from pounding the pavement daily. He was unable to run without experiencing a great deal of knee and leg pain and swelling. The knee pain was so severe, he was thinking of giving up running. He began taking proteolytic enzymes while continuing his regimen. Even though he maintained the workouts, the pain decreased. After a week's time, he was practically pain free. Within two weeks, he experienced no pain, though running seven miles per day on pavement, plus running up and down hills. Ultimately, he helped his high school team take third at state in cross country.

Dr. Baumuller[11] investigated the application of enzymes in sports-related ankle joint injuries, using a double-blind study. Treatment with ice and tape was used concurrently in all cases. The enzyme group experienced faster reduction of swelling, faster return to joint mobility, and decreased pain during resting or movement as compared to the placebo group. The time during which patients were unable to work or train was significantly shorter (reduced by some 50 percent) with enzyme therapy.

George was a college professor and soccer coach. On November 17th, after demonstrating a soccer kick, he felt a sharp pain in his lower back, radiating down into his right leg, particularly severe in the calf. He began taking proteolytic enzymes and experienced rapid improvement.

X rays indicated a decrease in the intervertebral disc space at L-5/S-1 and a bilateral spondylolytic defect (separation in the bone) at L-5.

According to the orthopedic surgeon who evaluated him some nine weeks after the injury on January 28th, there was little doubt this man had suffered a definite lumbar disc injury (slipped disc) with right leg nerve pain. The orthopedist felt George had made a remarkably good recovery and was essentially asymptomatic at the time of examination. The doctor was amazed at his swift recovery and concluded the patient required no further medical treatment.

Preventing Injury

In general, proteolytic enzymes taken prophylactically offer several advantages to sports participants exposed to injuries (boxing, wrestling, football, baseball, etc.). Also bursitis, tendosynovitis, tennis elbow, etc., are prevented or helped by it. These types of injuries are often difficult to treat.

Some doctors and researchers feel that enzymes taken orally are far safer than injectibles and do not have serious side effects because the body takes longer to assimilate the oral enzymes as opposed to injected enzymes. They enter the bloodstream and cells over a longer duration and do not expose an area to such a concentrated amount in a short time period.

Remember that back pain you've been carrying around like some unwanted baggage? Enzymes could be a welcomed adjunct to your present therapy.

Enzymes and Injury

1. **Take enzymes as early as possible.**
   It has been established that the therapeutic effects of enzymes are greater the earlier they are employed. Prophylactic use of enzymes is suggested for everyone, but especially for athletes involved in martial arts or contact sports. Further, Dr. Kleine reports that the duration of injuries and inability to train are both appreciably reduced if enzymes are taken before competitions.[12]

2. **Higher initial doses.**
   Enzyme therapy should always be started with relatively high doses. This is because intestinal bacteria are continually producing inhibitory substances which have to be eliminated before enzymes can be absorbed.

3. **Keep a safety margin before and after meals.**
   If enzymes are taken to treat injury (rather than as digestive aids), it is best to take them one-half hour before, or one and a half hours after meals. If not, there is danger that the enzymes will be inactivated by any inhibitors in the foods (especially with peas, lentils, or beans).

4. **No worries about the large numbers of tablets.**
   If patients find it difficult to swallow pills, enzymes are also available as granules or in sublingual forms. The granules can be stirred into foods such as yogurt or applesauce.

Nontraumatically Caused Inflammation

The inflammatory reaction is part of every injury, but it is involved in the majority of illnesses as well. These conditions will not be discussed in this book. Examples would include: acute and chronic bronchitis, sinusitis, cystitis and lower urinary tract infections (URI), pelvic inflammatory disease (PID), and even the common cold. Proteolytic enzymes help fight these conditions.

Corticosteroids: The Double-Edged Sword

Corticosteroids (including cortisone and hydrocortisone) are frequently prescribed anti-inflammatories and are used to treat a variety of conditions which involve inflammation. Though helpful in such conditions as rheumatoid arthritis, systemic lupus erythematosus, bronchial asthma, and so forth, their long-term use should be weighed against their undesirable effects. This could be a double-edged sword.

CORTICOSTEROIDS

Affect carbohydrate metabolism—and can lead to glucose intolerance
Affect protein metabolism—and can lead to severe muscle-wasting
Affect fat metabolism—and can cause redistribution of fat
Affect electrolyte and water metabolism—leading to sodium and water retention
Suppress immunity—this is why they're given to tissue transplant patients
Can cause gastric or duodenal ulcers
Can cause congestive heart failure (after high doses)
Can cause excitability, convulsions or hallucinations

Alternative Courses of Action

In recent years, researchers have offered an alternative to cortisone therapy with its serious and long-term side effects: the use of systemic enzyme therapy.

Those patients suffering from any injury (sports, on-the-job, or home) now have a safe alternative to cortisone therapy. This could lead to a reduction in drug-oriented side effects for the patient.

Acute, subacute, and chronic inflammations where enzymes might be helpful include:

Prophylaxis and therapy of postoperative inflammations and swellings
Sports injuries and prophylaxis

Lumbar disc conditions
Fractures, contusions, strains, tendinitis, tenovaginitis, bursitis, and subluxations
Varices, varicose phlebitis, phlebitis of the upper and inner veins, thrombophlebitis, phlebothrombosis
Post-thrombotic syndrome (leg, arm, pelvis)
Phlebectomy
Acute and chronic bronchitis, putrid bronchitis, bronchiolitis, bronchopneumonia, pneumonia, empyema, pulmonary abscesses, dry and exudative pleuritis, bronchial asthma
Metritis, parametritis, adnexitis, cystitis, urethritis, orchitis, epididymitis, sinusitis, pharyngitis, laryngitis
Angina pectoris

In a Nutshell

Certain enzymes (particularly proteolytic enzymes) shorten the inflammatory process in two ways: They reduce the duration of inflammation and they stimulate the body's own natural enzymatic activity without suppressing the immune system. These features alone present a considerable advantage over the use of corticosteroids.

In addition to this, enzyme therapy has been shown to:

Accelerate the healing of wounds
Eliminate pain
Avoid disfiguring scars
Have a fibrinolytic effect without hemorrhagic tendency
Increase the permeability of tissues
Shorten the duration of the disease
Be well-tolerated
Be free of undesirable side effects

We probably can't prevent each and every injury through better conditioning, but with enzymes acting as safeguards, we can reduce the severity of injury, heal faster, and return to our active lifestyles sooner in the event of injury. Finally, institute the Five-Step Jump-Start Plus Enzyme Program.

References

1. M. G. Cirelli, "Inflammation and Edema," *Medical Times* 92:919–922 (1964).
2. M. Masson, "Bromelain in Blunt Injuries of the Locomotor System. A Study of Observed Applications in General Practice," *Forschr Med* 113(19):303–306 (10 July 1995).
3. E. A. Fulgrave, "Enzyme Therapy and Sports Injuries," *Annals of the New York Academy of Science* 68:192 (1957).
4. Michael W. Kleine, M.D., "Therapie des Herpes Zoster mit Proteolytischen Enzymen," *Therapiewoche* 37:1108–1112 (1987).
5. Dr. W. F. (Rob) Rathgeber, "The Use of Proteolytic Enzymes (Chymoral) in Sporting Injuries," *South African Medical Journal* 45(S):181–183 (1971).
6. Henry T. Holt, "*Carica papaya* as Ancillary Therapy for Athletic Injuries," *Current Therapeutic Research* 11(10):621–624 (1969).
7. Anthony Cichoke, D.C., and Leo Marty, M.S., "The Use of Proteolytic Enzymes with Soft Tissue Athletic Injuries," *The American Chiropractor* 32-33 (Sept./Oct. 1981).
8. Hans-Dieter Rahn, "Efficacy of Hydrolytic Enzymes in Surgery," (Symposium on enzyme therapy in sports injuries) XXIV FIMS World Congress of Sport Medicine. Amsterdam: Elsevier Science Publishers, May 29, 1990, 5–8.
9. A. L. Lichtman, "Traumatic Injury in Athletes," *International Rec. Medicine* 170:322–325 (1957).
10. Max Wolf and Karl Ransberger, *Enzyme Therapy*, Los Angeles: Regent House, 1972, 94.
11. Marcel Baumuller, "Therapy of Ankle Joint Distortions with Hydrolytic Enzymes; Results of Double-Blind Clinical Trials," *Allgemeinmedizin* 19:178–182 (1990).
12. Michael W. Kleine, M.D. Presentation, "Systemic Enzyme Therapy in Traumatology," 13th Symposium in Lindau, 17 Nov., 1990.

CHAPTER 12

Chronic Condition Buster and Immune System Booster

WHAT ABOUT chronic problems? Lost a step or two? Are your joints continually stiff and aching? Can't climb stairs or grip a fork the way you used to? Does pain stay with you, coming and going for months and years? If so, you probably have a chronic disorder and enzymes might be your answer.

According to Webster's, chronic means "marked by a long duration or frequent recurrence." These are diseases that are continual or come and go, without ever going away completely. Arthritis, multiple sclerosis, HIV/AIDS, cancer, and herpes zoster are all examples of chronic disorders. Chronic disorders and their resulting pain are depressing and highly frustrating to both doctor and patient because modern medicine hasn't unlocked the mystery of their cure.

Two major factors in chronic disorders are *inflammation* and the *immune system.*

Acute inflammation is a series of processes (including reaction, repair, and renovation of the damaged area) essential for recovery. However, chronic inflammation is more complicated and usually involves some pathological process plus a weakened immune system.

THE IMMUNE SYSTEM

The second major factor in chronic disorders is the *immune system.* By reinforcing the immune system, our body is able to maintain health and to heal us when we are sick. Medicine doesn't cure, but can only contribute to our healing by relieving the strain on our bodies. Healing and maintenance of health are the responsibility of the body's own defenses.

The immune system is a complex network of specialized cells and organs which defends us against attacks by foreign invaders. When functioning properly, it fights off infections by bacteria, viruses, fungi and parasites. When it malfunctions, however, it can unleash a torrent of chronic diseases, from allergy to arthritis, and cancer to AIDS.

The immune system evolved because we live in a sea of microbes. Like man, these organisms are programmed to perpetuate themselves. The human body provides an ideal habitat for many of them, so they try to invade. Because the presence of these organisms is often harmful, the body's immune system will attempt to bar their entry or, failing that, seek out and destroy them.

The complex immune system displays several remarkable characteristics. It can distinguish between *self* and *nonself.* It is able to remember previous experiences and react accordingly. For instance, once you have had chicken pox, your immune system will prevent you from getting it again. The immune system is not only able to recognize millions of distinctive nonself molecules, but it can also produce molecules and cells to match and counteract every one of them. It has a sophisticated array of weapons at its command.

The success of the immune system relies on an incredibly elaborate communications network. Millions and millions of cells, organized into sets and subsets, pass information back and forth like clouds of bees swarming around a hive. The result is a sensitive system of checks and balances that produces a prompt immune response that is effective, appropriate, and self-limiting.

Self and Nonself

At the heart of the immune system is the ability to distinguish between self and nonself. Virtually every body cell carries distinctive molecules that identify it as self.

The body's immune defenses do not normally attack tissues that carry a self-marker. Instead, immune cells and other body cells coexist peacefully in a state known as self-tolerance. But when immune defenders encounter cells or organisms carrying molecules that say foreign, the immune troops move quickly to eliminate the intruders.

Any substance capable of triggering an immune response is called an *antigen.* An antigen can be a virus, a bacterium, a fungus or a parasite, or even a portion or product of one of these organisms. Tissues or cells from another individual (except an identical twin whose cells carry identical self-markers) also act as antigens. Because the immune system recognizes transplanted tissues as foreign, it rejects them. This is why transplant patients take immunosuppressive drugs—to suppress the immune system so it will accept these foreign tissues. The body will even reject nourishing proteins unless they are first broken down by the digestive system into their primary, nonantigenic building blocks.

In abnormal situations, the immune system can wrongly identify self as nonself and execute a misdirected immune attack. This results in an autoimmune disease such as rheumatoid arthritis or systemic lupus erythematosus.

In some people, universal substances such as ragweed pollen or animal dander can provoke the immune system to set off the inappropriate and harmful response known as allergy; in these cases, the antigens are known as *allergens.*

The Anatomy of the Immune System

The organs of the immune system are stationed throughout the body. They are generally referred to as *lymphoid organs* because they are concerned with the growth, development and deployment of *lymphocytes,* the white cells that are the key operatives of the immune system. Lymphocytes, which number about a trillion, bear the major responsibility for carrying out the activities of the immune system.

Lymphoid organs include the bone marrow and the thymus, as well as lymph nodes, spleen, tonsils and adenoids, the appendix, and Peyer's patches (clumps of lymphoid tissue in the small intestine). The blood and lymphatic vessels that carry lymphocytes to and from the other structures can also be considered lymphoid organs.

Cells destined to become immune cells, like all other blood cells, are produced in the bone marrow (the soft tissue in the hollow shafts of long bones). The descendants of some so-called stem cells become lymphocytes, while others develop into a second major group of immune cells typified by the large cell- and particle-devouring white cells known as *phagocytes.*

The two major classes of lymphocytes are *B cells* and *T cells.* B cells complete their maturation in the bone marrow. Each B cell is programmed to make one specific antibody. For example, one B cell will make an antibody which blocks a virus that causes the common cold, while another produces antibodies that zero in on the bacterium which causes pneumonia.

T cells, on the other hand, migrate to the thymus, a multilobed organ that lies high behind the breastbone. There they multiply and mature into cells capable of producing an immune response; that is, they become immunocompetent. In a process referred to as "T cell education," T cells in the thymus learn to distinguish self cells from nonself cells; T cells that would react against self antigens are eliminated.

T cells contribute to the immune defenses in two major ways. Regulatory T cells are vital to orchestrating the elaborate system. B cells, for instance, cannot make antibodies against most substances without T cell help. Cytotoxic T cells, on the other hand, directly attack body cells that are infected or malignant.

Chief among the regulatory T cells are "helper/inducer" cells. Typically identifiable by the T4 cell marker, helper T cells are essential for activating B cells and other T cells as well as natural killer cells and macrophages (the PacMen of the body). Another subset of T cells acts to turn off or suppress these cells.

They can also produce anaphylactic shock, a life-threatening allergic reaction characterized by swelling of body tissues, including the throat, and a sudden fall in blood pressure.

Upon exiting the bone marrow and thymus, some lymphocytes congregate in immune organs or lymph nodes. Others—both B and

T cells—travel widely and continuously throughout the body. They use the blood circulation as well as a bodywide network of lymphatic vessels similar to blood vessels.

Laced along the lymphatic routes—with clusters in the neck, armpits, abdomen, and groin—are small, bean-shaped lymph nodes. Each lymph node contains specialized compartments that house platoons of B lymphocytes, T lymphocytes, and other cells capable of enmeshing antigens and presenting them to T cells. Thus, the lymph node brings together the several components needed to spark an immune response.

The spleen, too, provides a meeting ground for immune defenses. A fist-sized organ at the upper left of the abdomen, the spleen contains two main types of tissue: the red pulp, where worn-out blood cells are eliminated, and the white pulp, which contains lymphoid tissue. Like the lymph nodes, the spleen's lymphoid tissue is subdivided into compartments that specialize in different kinds of immune cells. Microorganisms carried by the blood into the red pulp become trapped by the immune cells known as *macrophages* (large and versatile immune cells which act as microbe-devouring phagocytes, antigen-presenting cells, and are an important source of immune secretions). Although people can live without a spleen, persons whose spleens have been damaged by trauma or by disease such as sickle cell anemia are highly susceptible to infection; surgical removal of the spleen is especially dangerous for young children and the immunosuppressed.

Nonencapsulated clusters of lymphoid tissue are found in many parts of the body. They are common around the mucous membranes lining the respiratory and digestive tracts, areas that serve as gateways to the body. They include the tonsils and adenoids, the appendix and Peyer's patches.

The lymphatic vessels carry lymph, a clear fluid that bathes the body's tissues. Lymph, along with the many cells and particles it carries (notably lymphocytes, macrophages, and foreign antigens), drains out of tissues and seeps across the thin walls of tiny lymphatic vessels. The vessels transport the mix to lymph nodes, where antigens can be filtered out and presented to immune cells.

Additional lymphocytes reach the lymph nodes and other immune tissues through the bloodstream. Each node is supplied by an artery and a vein; lymphocytes enter the node by traversing the walls of very small specialized veins.

All lymphocytes exit lymph nodes via outgoing lymphatic vessels. Much as small creeks and streams empty into larger rivers, the lymphatics feed into larger and larger channels. At the base of the neck, large lymphatic vessels merge into the thoracic duct, which empties its contents into the bloodstream.

Once in the bloodstream, the lymphocytes and other assorted immune cells are transported to tissues throughout the body. They patrol everywhere for foreign antigens, then gradually drift back into the lymphatic vessels, to begin the cycle all over again.

The Cells and Secretions of the Immune System

The immune system stockpiles a tremendous arsenal of cells. Some staff the general defenses, while others are trained on highly specific targets. To work effectively, however, most immune cells require the active cooperation of their fellows. Sometimes they communicate through direct physical contact, sometimes by releasing versatile chemical messengers.

In order to have room for enough cells to match millions of possible foreign invaders, the immune system stores just a few of each specificity. When an antigen appears, those few specifically matched cells are stimulated to multiply into a full-scale army. Later, to prevent this army from overexpanding wildly, like a cancer, powerful suppresser mechanisms come into play.

Immune Complex Diseases

Immune complexes (IC) are clusters of interlocking antigens and antibodies. Under normal conditions, immune complexes are rapidly removed from the bloodstream by macrophages in the spleen and Kupffer's cells in the liver. In some circumstances, however, immune complexes continue to circulate. Eventually they become trapped in the tissues of the kidneys, lung, skin, joints or blood vessels. Just where they end up probably depends on the nature of the antigen, the class of antibody (IgG, for instance, instead of IgM) and the size of the complex. There they set off reactions that lead to inflammation and tissue damage.

Immune complexes work their damage in many diseases. Sometimes, as is the case with malaria and viral hepatitis, persistent low-grade infections occur. Sometimes they arise in response to environmental antigens, such as the moldy hay that causes the disease known as farmer's lung. Frequently, IC develop in autoimmune disease where the continuous production of autoantibodies overloads the immune complex removal system.[7]

Autoimmune Disease

Sometimes the immune system's recognition apparatus breaks down, and the body begins to manufacture antibodies and T cells directed against the body's own constituents—cells, cell components, or specific organs. The body attacks itself. Such antibodies are known as *autoantibodies*, and the diseases they produce are called *autoimmune diseases* (not all autoantibodies are harmful; some types appear to be integral to the immune system's regulatory scheme).

Autoimmune reactions contribute to many enigmatic diseases. For instance, autoantibodies to red blood cells can cause anemia; autoantibodies to pancreas cells contribute to juvenile diabetes; and autoantibodies to nerve and muscle cells are found in patients with the chronic muscle weakness known as myasthenia gravis. The autoantibody known as rheumatoid factor is common in persons with rheumatoid arthritis. And then there is the autoimmune disease known as AIDS.

These disorders can usually be divided into *organ-specific* and *organ nonspecific* auto-aggressive diseases. Please see Table 12–1.

Persons with systemic lupus erythematosus (SLE), whose symptoms encompass many systems, have antibodies to many types of cells and cellular components. These include antibodies directed against substances found in the cell's nucleus—DNA, RNA, or proteins—which are known as antinuclear antibodies, or ANAs. These antibodies can cause serious damage when they link up with self antigens to form *circulating immune complexes* (CIC), which become lodged in body tissues and set off inflammatory reactions.

Table 12-1. Auto-Aggressive Diseases

Organ-specific disorders:
- a. Thyroiditis
- b. Myasthenia gravis
- c. Goodpasture's syndrome
- d. Pemphigus
- e. Multiple sclerosis
- f. Diabetes mellitus (Type 1)

Organ nonspecific disorders:
- a. Chronic hepatitis
- b. Ulcerative colitis
- c. Rhematoid arthritis
- d. Lupus erythematosus

Normally the immune system is able to engulf immune complexes and break them down, enzymatically. However, this doesn't always happen. The IC sometimes alarms another death squad, called the *complement system.* Complement is the name given to a group of at least nine different protein-degrading enzymes.

The immune complexes always alarm the first complement in the system. The complement appears, attaches itself to the antibody, and calls for the second complement, then the third, and so on. One member of the system activates the next member. The whole sequence is similar to a row of upright dominoes which knock each other over in turn.

It is the last member, the ninth complement, that is the actual killer. This complement breaks down the immune complex.

The reason for this series of activations is safety. The body does all it can to avoid activating that dangerous last killer at the wrong time and in the wrong place, because it could conceivably destroy the entire body. This mechanism is a lot like the fail-safe system we use to protect against accidental atomic bomb activation. It's not possible to simply push one button; several consecutive, preliminary steps must first be taken before anything can be activated.

In theory, this is great. However, in practice, sometimes things go wrong. For example, the activation of macrophages and complement system depends, among other things, on the size and numbers of the immune complexes and whether they are floating freely in the blood or lymph or are deposited in the tissues.

If immune complexes deposit themselves in the tissues, the complement system will arrive and bring about an inflammatory response, which, in turn leads to tissue destruction. This is the beginning of an autoimmune disease. For example, if immune complexes collect in the kidneys, complement activation causes inflammation which leads to *glomerulonephritis.*

Research has shown that certain enzyme mixtures, by interrupting the complement cascade, are able to prevent the glomerulonephritis.

Improvement in several IC-related diseases (including Crohn's disease,[1-3] ulcerative colitis, [1-3] hepatitis,[4] rheumatoid arthritis,[5-7] multiple sclerosis,[8] certain tumors,[9-11] pancreatitis,[12] systemic lupus erythematosus[13] and glomerulonephritis[14]) has been associated with the elimination of IC. Plasmapheresis or protein A immunoadsorption has been used to reduce circulating immune complex levels.[15,16] Reduction of the circulating immune complexes has also been achieved by treatment with hydrolytic enzymes, especially combinations of proteolytic, lipolytic, and amylolytic enzymes.[17-20] These enzymes mobilize cell-bound IC (call out the infantry) and stimulate macrophages to clear the intruders out.[21] Furthermore, hydrolytic enzymes prevent activation of the complement system and cell destruction by splitting the part of the cell-bound immune complexes.

Other disorders with similar processes can be treated using enzyme mixtures. These include Crohn's disease or ulcerative colitis (the chronic intestinal inflammation caused indirectly by the deposition of immune complexes in the intestinal tissue).

Enzyme therapy can cause interruption of the self-damaging complement cascade by breaking up pathogenic immune complexes and by activation of the macrophages. This breaks the vicious circle which otherwise results in continual deterioration of the involved tissue and is the cause for many chronic disorders. Through the use of enzyme therapy, the never-ending circle is broken.

The list of autoaggressive disorders caused by immune complexes is long and affects many organs. In the lungs, it can lead to pulmonary fibrosis, in the pancreas, chronic recurring pancreatitis. Rheumatic diseases, for example, chronic rheumatoid arthritis or ankylosing spondylitis, and inflammatory diseases of the intestines (ulcerative colitis or Crohn's disease) are typical autoaggressive disorders.

Autoimmune diseases affect the immune system at several levels. In patients with systemic lupus erythematosus, for instance, B cells are hyperactive while suppresser cells have decreased activity. It is not clear which defect comes first. Patients with rheumatoid arthritis, who have a defective suppresser T-cell system, continue to make antibodies to a common virus, whereas the response normally shuts down after about a dozen days.

No one knows just what causes an autoimmune disease, but several factors are likely to be involved. These may include viruses and environmental factors such as exposure to sunlight, certain chemicals, and some drugs, all of which may damage or alter body cells so that they are no longer recognizable as self. Sex hormones may be important too, since most autoimmune diseases are far more common in women than in men. Heredity also appears to play a role.

Many types of therapies are being used to combat autoimmune diseases. These include corticosteroids, immunosuppressive drugs developed as anticancer agents, radiation of the lymph nodes, plasmapheresis (a sort of "blood washing" that removes diseased cells and harmful molecules from the circulation), and enzyme therapy.

Enzyme therapy is able to break down and remove pathogenic immune complexes, to stimulate the body's endogenous defenses, and to accelerate the mechanisms of inflammation, thus resolving the condition more quickly.

An initial deterioration in the tissue can occur. The enzyme mixtures are able to break up the immune complexes deposited in the tissues to smaller sizes, and thus, bring them back into the bloodstream. The increased presence of immune complexes in the bloodstream can temporarily cause an increase in the severity of the disease symptoms. However, the enzymes (if introduced into the body in sufficient quantities) can soon cope with the immune complexes in the blood and the disease symptoms should subside.

The initial disease symptom effects have been studied extensively by Professor Steffen of the Institute for Immunology, University of Vienna.[22] Rabbits with pathogenic immune complexes were treated with a concentrated enzyme solution. Professor Steffen found that certain enzyme mixtures have particular application with inflammatory diseases. Furthermore, he found that the more concentrated the enzyme mixture, the more the immune complexes were destroyed. Finally, all immune complexes were broken down within a few hours and the rabbits returned to health.

It also was observed that the rate of IC degradation was reduced when they contained particularly large quantities of antibodies. It was found that the enzymes degraded the ICs as a result of their action on the antibodies.

If large quantities of antibodies were present in the immune complexes, large quantities were degraded. The result of this action was an increase in the numbers of immune complexes. However, this was beneficial because they were small in size, allowing gradual and complete degradation.[22]

Enzymes have a beneficial effect on the immune system.

It is known that physical stress increases the numbers of immune fighter cells, for example, T- and B-lymphocytes.

Dr. Wolfgang Bringmann of Berlin, Germany, decided to test forty athletes at an official mini-marathon.[23] Beginning two days before the event, the athletes took ten coated tablets of enzymes, three times per day. On the morning of the event, those taking part in the study again took ten coated tablets of the enzyme preparation.

The test subjects were trained athletes and no training schedule was interrupted for the test. The immunological parameters were taken from blood samples taken from the athletes by Dr. Rudolf Kunze, also of Berlin, and produced interesting results.

Dr. Kunze found that this increase of lymphocytes was appreciably reduced when enzymes were administered to the athletes. Put simply, the athlete's body was less stressed.

As every sports doctor knows, highly trained athletes are very susceptible to infections. This is probably associated with the fact that sports activities reduce immune defenses during athletic stress. The reason for this "downregulation" has been postulated as being the sport or stress-dependent increase in blood ß-endorphin levels.

Dr. Kunze found that certain enzyme mixtures are able to strengthen the immune system and its ability to respond to infection.[23]

Enzymes can help athletes compete in highly stressful sports. By taking enzymes from the beginning of training, female skiers can avoid the elevated creatine kinase and myoglobin levels that usually occur under physical stress. In addition, these female athletes who take the enzymes experience less muscle pain and stiffness and train with fewer health problems.

Enzyme Therapy

Enzyme therapy is used both locally and systemically. Locally, it can serve as a substitute for enzymes in pancreatic insufficiency; in thrombolysis; chemonucleolysis (with chymopapain); in debridement of ulcers; or as an ointment in burn dressings. Systemically, enzyme therapy is used in the treatment of arthritis, AIDS, cancer, cardiovascular disorders, multiple sclerosis and other chronic conditions.

As of 1990, an estimated 1.5 million Americans were infected with the human immunodeficiency virus (HIV), according to the Centers for Disease Control in Atlanta. Research in Germany and Puerto Rico has shown positive results in battling this dreaded disease with enzyme therapy.

Enzyme therapy has also been shown to be effective in treating multiple sclerosis (MS). Victims of MS must live with the knowledge that their condition will probably worsen. Enzyme therapy, however, might offer new hope because it can reduce the severity of symptoms.

Rheumatism and multiple sclerosis are quite similar in many ways. Joint rheumatism is usually treated with immunosuppressive or inflammation-inhibiting drugs. As with MS, these methods bring limited success plus serious side effects. Rheumatoid arthritis improves or is at least delayed when high doses of enzymes are used over an extended period of time.

Nearly 900,000 Americans die from circulatory diseases every year. This is the number-one killer in our country. Enzyme therapy might be the answer as it activates the fibrinolytic system and stimulates beneficial cells (such as macrophages) without the long-term side effects of drug therapy.

What about cancer? At the turn of the twentieth century, John Beard, a British embryologist, began to treat cancer patients with enzymes. Since that time a growing number of researchers and physicians have found the beneficial effects of treating cancer with enzyme therapy.

In Review

With skyrocketing health care costs, we, the public, must have access to *everything* that can improve our health. Enzymes, through uncooked foods and supplements, can help every system in your body function better. The better your body works, the better you'll feel.

Enzymes could be used to alleviate much pain and suffering, as a protection against deadly disease, thus allowing longer, healthier and happier lives. Frequently, the very people suffering from pain and disease who need enzymes are not aware of them. Many doctors, inundated with volumes of literature, know very little, if anything, of enzyme benefits. It is vital that each individual acquire some personal knowledge of enzymes and their functions. This knowledge and use of enzymes can help us take charge of our own lives.

Chronic disease is rampant in the United States. Enzyme therapy effectively helps fight chronic disease, reduces the severity of injuries, reduces the time lost from injuries, aids digestion and slows down aging.

We often hear about miracle drugs, revolutionary new surgeries, plus vitamins, minerals and herbs to heal mankind—but what about enzymes? They are the very basis of our existence. Be selfish about your health and that of your loved ones—jump-start your day the enzyme way.

References

1. R. Fiasse, A. Z. Lurhuma, et al., "Circulating Immune Complexes and Disease Activity in Crohn's Disease," *Gut* 19:611–617 (1978).
2. I. Kre, Z. Kojecky, I. Matouskova, and L. Benysek, "Crohn's Disease, Serum Immunodepressive Factors and Circulating Immune Complexes," *Bollettino Dell Istituto Sieroterapico Milanese* 59:619–624 (1980).
3. H. J. F. Hodgson, B. J. Potter, and D. P. Jewell, "Immune Complexes in Ulcerative Colitis and Crohn's Disease," *Clinical and Experimental Immunology* 29:187–196 (1977).

4. A. Musca, C. Cordova, et al., "Circulating HBsAg/IgM Complexes in Acute and Chronic Hepatitis B," *Hepato-Gastroenterology* 31:208–210 (1984).
5. B. R. Kaye, "Rheumatologic Manifestations of Infection with Human Immunodeficiency Virus (HIV)," *Annals of Internal Medicine* 111:158–167 (1989).
6. G. Klein, H. Schwann, and W. Kullich, "Enzymtherapie bei Chronischer Polyarthritis," *Natur. Ganzheitsmed.* 1:112–116 (1988).
7. K. Fehr, "Die Bedeutung von Immunprozessen in der Pathogenese Entzundlich-Rheumatischer Erkrankungen," *Akt. Rheumatol.* 9:1–12 (1984).
8. M. K. Dasgupta, G. W. Kenneth, V. Kaiviayil, and J. B. Dossetor, "Circulating Immune Complexes in Multiple Sclerosis; Relation with Disease Activity," *Neurology* 32:1000–1004 (1982).
9. H. Jager, M. Popescu, and W. Kaboth, "Circulating Immune Complexes in HIV Infection," 2nd Int. Symp. Immunobiol. in Clin. Oncology and Immune Dysfunctions, Nice, April 4–7, 1987.
10. U. E. Nydegger, "Biological Properties and Detection of Immune Complexes in Animal and Human Pathology," *Reviews of Physiology Biochemistry and Pharmacology* 85:64–111 (1979).
11. K. E. Hellstrom, I. Hellstrom, et al., "Blocking (Suppressor) Factors, Immune Complexes and Extracorporeal Immunoadsorption in Tumor Immunity," In: Salinas, A. F., Hanna, G. M. (eds) *Immune Complexes and Human Cancer*, Vol. 15. New York: Plenum Press, 1985, 213–230.
12. A. N. Theofilopoulos, "Evaluation and Clinical Significance of Circulating Immune Complexes," *Progress in Clinical Immunology* 4:63–106 (1980).
13. W. Samtleben and H. J. Gurland, "Plasmapherese bei Lupusnephritis; Rationale Basis und Klinische Erfahrungen," *Nieren Hochdruckkr.* 3:104–108 (1989).
14. K. J. Johnson and A. W. Peter, "Newer Concepts in the Pathogenesis of Immune Complex-Induced Tissue Injury," *Laboratory Investigation* 3:218 (1982).
15. H. Jager, "Hydrolytic Enzymes in the Treatment of Patients with HIV-Infections," from lecture, First International Conference on Systemic Enzyme Therapy. September 11, 1990.
16. D. D. Kiprov, R. Lippert, et al., "The Use of Plasmapheresis, Lymphocytapheresis, and Staph Protein-A Immunoadsorption as an Immunomodulatory Therapy in Patients with AIDS and AIDS-Related Conditions," *Journal of Clinical Apheresis* 3: 133–139 (1986).
17. G. Stauder, D. Fuchs, et al., "Adjuvant Therapy of HIV Infections with Hydrolytic Enzymes: Course of Heopterin, CD4-T-Cells, Immune Complexes (IC), and Clinical Efficacy," 8th International Workshop on Biochemical and Clinical Aspects of Peridines, St. Christoph (Tyrol), February 11–18, 1989.

18. G. Stauder, "Hydrolytische Enzyme als Adjuvante Therapie bei HIV Infektionen," Munchener AIDS-Tage 1990: *Abstraktband*, S 36–37 (1990).
19. Karl Ransberger, W. Van Schaik, W. Pollinger, and G. Stauder, "Naturheilkundliche Therapie von AIDs mit Enzympraparaten," *Forum des Praktischen und Allgemeinarztes* 27. Heft 4 (1988).
20. G. Stauder, K. Ransberger, et al., "The Use of Hydrolytic Enzymes as Adjuvant Therapy in AIDS/ARC/LAS Patients," *Biomedecine et Pharmacotherapie* 42: 31–34 (1988).
21. R. J. Bonney and P. Davies, "Possible Autoregulatory Functions of the Secretory Products of Mononuclear Phagocytes," in: D. O. Adams, G. M. Hanna (eds) *Macrophage Activation*, vol. 13. New York: Plenum Press, 1984, 198–219.
22. C. Steffen and J. Menzel, "Enzymbbau von Immunkomplexen," *Zeitschrift fur Rheumatologie* 42:249–255 (1983).
23. Rudolph Kunze, Presentation at 13th Symposium in Lindau, 17 Nov., 1990—Systemic Enzyme Therapy in Traumatology.

CHAPTER 13

More Precious Than Gold

MARGARET IS ninety-three years old. Her feet and fingers are gnarled and deformed. She has stiff, swollen, and tender joints, muscular aches and pains. X rays reveal visible bony lipping and spurring of the joints. Her movement is restricted with gradual permanent fusion of the spine, leading to increased deformity and immobility. Her condition is known as arthritis.

George is a fifty-five-year-old man, riddled with rheumatoid arthritis, particularly in his hands and feet. For over ten years, he has repeatedly experienced the surgeon's knife in the painful and constant hope of a cure. Sad to say, the surgeries resulted in greater joint deformation and increased, constant pain. He is left hopeless and helpless.

At seventy, Ann has great difficulty walking. Her weight-bearing joints have worn down. This condition is called osteoarthritis (usually found in elderly people). Osteoarthritis develops as a result of continuous wearing away of the joint cartilage.

Drugs and surgery seem to be organized medicine's only answer to these conditions. The phrase, "Take two aspirins and call me in the morning," rings ad nauseam through the hallowed halls of traditional health care.

There must be more to fighting arthritis than a bin full of drugs and the surgeon's knife. There *is* more, and a combined program emphasizing enzyme therapy might be the answer.

AWFUL ACHING ARTHRITIS

The terms rheumatism, arthritis, joint disease, soft tissue disorders, etc., can represent an apparently limitless range of diseases that seem to fall under the arthritic or joint disease umbrella. These conditions range from problems of the spine (ankylosing spondylitis) to polyarthritis, from connective tissue disorders such as lupus to gout and the list keeps growing.

Arthritis, a painful swelling of the joints, is a major disease that disables people of all ages, but particularly the elderly. During movement, the ends of the bones are normally protected from wear by cartilage and small sacs of fluid that act as cushions. But with age, bones sometimes disintegrate and the joints become malformed and painful. The cause of arthritis is unknown, but it afflicts millions around the world. Similar to multiple sclerosis, it is characterized by periods of waxing and waning.

Though the cause of arthritis is unknown, there are many theories. In rheumatoid arthritis, as with certain autoimmune conditions, the immune system mistakenly attacks bone coverings as if they were invading tissues. Further, there may be a valid link between diet and arthritis. In some individuals, it is also possible that certain foods may stimulate the immune system to attack. Diet may worsen arthritis, since integrity of the immune system depends on adequate nutrition. In some people, for example, milk and milk products seem to aggravate the disease.

Though we don't know the cause of arthritis, doctors and enzymologists agree that the disease is associated with pathogenic immune complex formation. Free radicals are also implicated as contributing to this painful condition. The effects of rheumatoid arthritis could be lessened if detected and treated in the early stages.

The now famous fatty acid found in fish oil, EPA (a type of omega-3 fatty acid), is another nutrient link to arthritis. Research shows that the same diet recommended for a healthy heart, that is, low in saturated fat and high in EPA, helps reduce or prevent the joint inflammation which makes arthritis so painful. According to research theory, EPA probably interferes with the action of prostaglandins, the chemicals involved in inflammation.

One possible link between nutrition and arthritis could be the lipid peroxidation reaction. Inflammation and swelling cause lipid peroxidation of the membranes within joints. Vitamin E is effective against peroxidation. But since its role is preventive, not restorative, the fact that it has not improved active cases of arthritis is not surprising.

Overweight is a known connection between arthritis and nutrition. Maintaining a normal weight is important since the joints affected are often weight-bearing joints. Oddly, however, weight loss often relieves the worst of the arthritis pain in the hands—which are not weight-bearing.

Although American therapy seems to be limited to surgery and drugs (with their assured side effects), European researchers have found a light at the end of the arthritic tunnel. Enzyme therapy appears to be that light.

One way to differentiate rheumatic conditions could be as follows:

1. **Inflammatory rheumatic conditions:**
   This arthritic condition is primarily inflammatory rheumatic joint disease.
2. **Soft-tissue rheumatic conditions:**
   Extremely varied degenerative and inflammatory diseases relating to ligaments, tendons, muscles, and synovial sheaths.
3. **Degenerative joint disease** (osteoarthritis):
   Neither pathology nor etiology is related to inflammatory rheumatism.

The extent of inflammatory changes in the synovial membrane helps differentiate degenerative joint diseases from inflammatory rheumatic disorders.

Rheumatic Diseases

Rheumatoid arthritis is probably one of the most common chronic diseases (there are over 100 types of arthritis). At least forty-three million people in the United States suffer from some type of this condition, and women are three times more susceptible than men. The government estimates that by the year 2020, sixty million Americans will have arthritis of some kind or another.[1]

Rheumatoid and juvenile arthritis are an inflammatory arthritis that attacks the synovial membranes surrounding the lubricating fluid in the joints. The tissues and cartilage are destroyed in and around the joints, and often the bone surfaces. The body replaces this damaged tissue with scar tissue, causing the spaces between the joints to become narrow, to develop folds, and to fuse together. The entire body is affected instead of just one joint, as in osteoarthritis. Rheumatoid arthritis creates stiffness, swelling, fatigue, anemia, weight loss, fever, and often crippling pain. It often occurs in people under forty years of age, including young children. Juvenile rheumatoid arthritis affects 50,000 younger Americans (aged eighteen and under); six times as many girls are afflicted with juvenile rheumatoid arthritis as boys.

The onset of rheumatoid arthritis is often associated with physical or emotional stress; however, poor nutrition or bacterial infection may also be the trigger.

Pathological changes in rheumatoid arthritis include the thickening of the synovial membrane. As excessive synovial fluid builds up, the synovial capsule swells, articular cartilage is encapsulated (resulting in breakdown), and remissions and exacerbations occur.

As discussed in the immune system chapter, circulating immune complexes have a definite influence in rheumatoid arthritis. Frequently, the antigen-antibody complexes trigger the arthritis disease, resulting in a pronounced atrophy of the synovial membrane. Further, since the circulating immune complexes (CIC) only become pathogenic after binding to tissue, the severity of the illness is dependent on how well the antigen-antibody complexes bind locally and the amount of CIC concentrated in the affected area.

Professor Steffen and associates studied the value of enzyme therapy in treating rheumatoid arthritis patients. Of the 42 studied, 26 improved (61.9 percent), 13 had no change, and 3 others worsened.[2] An improvement was observed more frequently in patients without antigen-antibody complexes or who were CIC positive prior to treatment and became CIC negative during the enzyme therapy.

Investigations by Horger, Moro, and Van Shaik reveal that elevated CIC values can be rapidly reduced to normal levels (in two to six weeks) with enzyme therapy.[3]

Ankylosing Spondylitis (Bekhterev's Disease)

Therapy for rheumatic diseases is not easy, especially in ankylosing spondylitis or Bekhterev's disease. Any medical intervention has to be a long-term effort and requires a high degree of compliance by the patient. This is a major problem because patients with Bekhterev's disease are notoriously difficult to motivate.

The etiology of ankylosing spondylitis is still poorly understood though immune complexes seem to be involved. In fact, as many as 43 percent of patients have increased levels of circulating immune complexes.

Ankylosing spondylitis is an inflammatory disease, specifically relating to the spinal column and sacroiliac joints. The pathology is similar to rheumatoid arthritis. This ossification of the entire spinal column (so-called bamboo spine) occurs in the late stages of the disease. Antigen-antibody complexes are seen in a large proportion of the patients.

In many instances, a person with bamboo spine is extremely bent over and cannot straighten up.

The properties which make proteolytic enzymes attractive in treating this crippling disease include:

The good anti-inflammatory activity
The capability of splitting not only circulating, but also tissue-resident, immune complexes
The low rate of side effects
Compatibility with other drugs.

Professor Goebel used proteolytic enzymes in his six-month, double-blind, randomized study on the efficacy of proteolytic enzymes with Bekhterev's disease.[4] During the first week of his study, it was noticed that Indomethacin provided faster pain relief than the enzymes. Goebel attributed this to the more powerful analgesic component of the nonsteroidal antirheumatic agent. But it soon became clear that this effect waned before the end of the second month. In the third month of the study, the Indomethacin group reported *more* pain than the enzyme group. After six months, there was a significant difference between the two groups in favor of the enzyme therapy.

During the early phase of Goebel's study, the immune complex titers in both groups initially increased. However, from the fourth week onward, a steady decline was recorded in the enzyme group which continued for the subsequent five months. Under Indomethacin therapy, however, the levels of circulating immune complexes did not fall.

As enzymes are more effective, but only start to work after some delay, Goebel suggested a combination of enzyme therapy with analgesics for the first four weeks. Patients with Bekhterev's disease tend to be impatient and are not prepared to wait for results.

Soft-Tissue Rheumatism (Nonarticular Rheumatoid Arthritis)

The list of terms designated as nonarticular (soft tissue) rheumatism often includes extraarticular rheumatic symptoms. Bursitis, fibromyalgia, myofascial pain syndrome, periarthritis, tendovitis, and tendomyositis are a few included in the list. Causes for all of these periarticular diseases are inflammatory, functional, and degenerative processes. Usually, soft-tissue rheumatism is treated symptomatically with NSAIDs (nonsteroidal anti-inflammatory drugs) and possibly muscle relaxants. Unfortunately, there are frequent side effects.

Table 13-1. Clinical Syndromes Associated with Rheumatism of Soft Tissues

---

Muscular rheumatism
  Myosis
  Tendomyosis
Periarthropathy syndromes
  Shoulder
  Hip
  Knee
External humeral epicondylitis
Tendovaginitis
Tendopathy
Bursitis
Bursopathy
Generalized tendopathy

The symptoms of soft tissue rheumatism vary widely, though the principal symptoms are pain and restricted movement. Systemic enzyme therapy is a low-risk therapeutic approach which can also be used against this form of rheumatism.

Dr. Klaus Uffelmann of Gemundsen, Germany reported on an eight-week, double-blind study of 424 patients.[5] Proteolytic enzymes proved particularly effective in alleviating pain, improving mobility, and reducing soft tissue swelling and muscle stiffness.

Two weeks after the study's completion, the researchers tested the participants and found that the improvement continued even after termination of the therapy. According to Dr. Uffelmann, this makes enzyme therapy an appropriate basic tool against arthritis.

Scientific publications and investigations (Medical Enzyme Research Association) acknowledge that enzymes are not only virtually free of side effects, but are as effective as the traditionally-prescribed drugs.

A study was published in 1983 involving 1,004 rheumatic patients and enzyme treatment.[4] The study involved 141 doctors. The rheumatic group contained a wide variety of cases including activated arthrosis, soft tissue rheumatism, basic rheumatism of the joints, and many other rheumatic disabilities. Case histories were analyzed and the patients were classified by particular rheumatic disorder. Of the patients, 76 to 96 percent were "improved" or "considerably improved," while 10 to 24 percent remained "unchanged." In only 2 percent of the patients was there a deterioration in their condition, and over 99 percent reported they were free of side effects.

Research at the Institute for Immunology in Vienna, Austria, the Rheumatism Clinic in Wiesbaden, Germany, and the Rheumatism Clinic in Bad Wiessee, Germany has positively confirmed the influence of enzymes on the immune system, including pain reduction, ease of movement, and reduced deterioration rate.

Degenerative Joint Disease (Osteoarthritis)

A degenerative joint disease (osteoarthritis) is related to the wear and tear of aging and involves deterioration of the cartilage at the ends of the bones. Friction results when the once smooth surface of cartilage becomes rough. The tendons, ligaments, and muscles holding the joint together become weaker, and the joint itself becomes deformed, painful, and stiff. There is little or no swelling, but usually some pain. Any disablement is usually minor.

Only rarely does osteoarthritis develop before the age of forty. Typically it's hereditary, but it afflicts almost three times as many women as men. Movement of the joint becomes increasingly painful and limited. Morning stiffness is typical. Symptoms may be alleviated with rest.

Osteoarthritis develops as a result of an unknown defect of the cartilaginous tissue. Progressively, there is thinning of the cartilaginous tissue. Later there are bulging prominences of the bone, cyst formation and the grinding of bone on bone.

A study of eighty patients suffering from osteoarthritis of the knee compared the effect of an enzyme mixture with diclofenac (a nonsteroidal anti-inflammatory and pain reliever frequently used in the treatment of arthritis). The researchers measured pain at rest, on motion, on walking, and at night, as well as pain tenderness. They found that the enzyme mixture was just as effective in treating the patients' pain as was diclofenac.[6]

Osteoarthrosis

An inflammation may develop from increased deposits of ground-down cartilaginous debris (for example, in the knee joint), with the resulting inflammatory reaction. This is referred to as osteoarthrosis. Enzyme therapy has been used successfully to treat osteoarthrosis in Germany for more than twenty years and is considered equivalent to diclofenac.[7]

Gout

Gout is the result of a metabolic error and occurs when there is too much uric acid in the blood, tissues, and urine. The uric acid crystallizes in the joints, acting as an abrasive and causing swelling and pain. Uric acid kidney stones may be a related problem. The majority of gout patients are male. The cardinal sign of gout is inflammation of the big toe.

Uric acid is a byproduct of certain foods, so this condition is closely related to diet. It can also be brought on by stress. Obesity and an improper diet increase the tendency for gout.

The term "gouty arthritis" is used primarily in relation to a sudden onset of excruciating pain, almost always in a peripheral joint, usually in a lower extremity. Most of the initial attacks involve the big toe, or, less frequently, the instep, ankle, or heel. Affected joints are red, hot, swollen, and tender—typical signs of inflammation.

Although there are no studies directly related to gouty arthritis, since there is inflammation of the joint, normalization of uric acid and inflammation-fighting proteolytic enzymes seem to be a logical treatment program.

Antirheumatic therapy

Patients with chronic polyarthritis also need a good deal of understanding, because the standard medical therapy only starts to work after several months. The only exception is the antineoplastic drug, Methotrexate. According to Dr. Gert Klein[8] of Saalfelden, Austria, today's drugs have the following special features:

The earlier therapy starts, the better is the effect
The drug will only start to work after several months of continuous use
There is a depot effect
All substances pose some risk to the patient.

Prescribed therapies are, unfortunately, quite limited. In the past decade, treatment of arthritis has become even more restricted as the serious, long-term effects of cortisone and other anti-inflammatory drugs have come to light. Steroids can quickly alleviate arthritis symptoms, but suppress the body's defenses, leading to complications more serious than the arthritis itself.

Nonsteroidal anti-inflammatory drugs (NSAIDs), on the other hand, alleviate the pain and symptoms of arthritis, but seem to have no effect on the joint degeneration associated with the disease.

The action of these drugs depends on the inhibition of prostaglandin. But prostaglandins are also needed in our bodies to protect the mucosa in the gastrointestinal tract from stomach acids. Therefore, the use of drugs that inhibit prostaglandin may cause bleeding, inflammation and ulcers.

The most widely used treatment for rheumatoid arthritis is gold. For hundreds of years, gold has been used to treat a multitude of conditions, including tuberculosis, leprosy, and syphilis.

Though it is still unclear exactly how gold works, its use has continued despite the fact that every third patient has suffered some side effect. These side effects can be extremely serious when gold is injected into the muscles. Additionally, gold builds up in the tissues over an extended period of time and acts as a poison, causing allergic skin reactions, anemia, and possibly, eye disease.

Penicillamine has been given to arthritis patients to counteract the side effects of gold therapy. Unfortunately, it is subject to the same risks as the gold therapy itself! Penicillamine has now simply become the scapegoat for gold therapy.

Dr. Bruckle of the Rheumatology Clinic, University of Basel, Switzerland, discussed gold therapy at a rheumatology training course and how it can affect patients.[4] According to Dr. Bruckle, the 40 percent success rate reported in earlier literature simply could not be proven under investigation. Actual figures are far lower. Arthritis symptoms (over a two-year period) improved in only one out of every six patients, and lasted an average of only ten months. Unfortunately, the longer you have arthritis, the less likely drugs will have a positive effect.

If all else fails and joint degeneration continues, the next step in therapy is prescription of cytostatic medications, the most common of which is Azathioprine®. Azathioprine hinders enzymatic activity in the immune system by inhibiting the production of the body's own defense and regenerative mechanisms. That is, it interferes with your jump-start enzymes. This treatment is usually abandoned because of its serious side effects, mainly disruption of blood production.

Dr. Klein, head of the Rehabilitation Center for Rheumatic Disorders and Cardiovascular Disease in Saalfelden, Germany, was not prepared to accept the risk of usual drug therapy in treating his arthritis patients.[9] He looked for an alternative approach and found proteolytic enzymes with their anti-inflammatory, fibrinolytic, immune complex-reducing and immunostimulating properties. Among the latter, Klein considers the increase in macrophage activity to be significant.

For a one-year period, twenty-four patients received coated tablets daily of either proteolytic enzymes (ten patients) or gold (fourteen patients). The groups did not differ with regard to the number of affected joints.

Improvements in symptoms were comparable in the two groups. According to Klein, results of this study suggest that the progress of the disease was checked by the enzyme therapy.

There were few side effects. In the enzyme group, gastrointestinal side effects were the most common (soft stools or diarrhea), while the gold group had one case of bleeding ventricular ulcer (the heart), one of exanthema (an eruptive skin disorder), and one of gingivitis (inflammation of the gums). The undesirable effects of the enzyme therapy can therefore be described as much more harmless than those which may be expected under gold therapy.

Dr. Klein conducted a second, six-month clinical study on rheumatoid arthritis patients and the effects of enzyme therapy.[7] Those in the enzyme group were long-time arthritis sufferers in advanced stages and were beyond the point at which gold was normally administered.

Usually, the longer a person has been suffering from rheumatoid arthritis, the less likely any treatment will be successful. Taking this into consideration, any sign of improvement with enzyme therapy would be thought of as a success.

Not only did the enzymes alleviate arthritic symptoms, but the treatment reduced external signs, such as joint swelling, pain and stiffness. The enzyme mixture also helped to degrade the fibrin (built up by tissue-immobilized immune complexes), which has been indirectly associated with the cause of rheumatoid arthritis. Through this action, the mechanisms which cause inflammation are quickly stopped and the possibility for greater deterioration is reduced.

But what impressed Dr. Klein more than anything else was that enzyme therapy caused fewer side effects than the gold therapy. Of the gold therapy patients, 20 percent complained of side effects, as compared to only 5 percent of the enzyme therapy group. This, in and of itself, should be sufficient proof of the overwhelming success of enzyme therapy.

There is one disadvantage to enzyme therapy—it has a delayed effect. Dramatic results will not happen overnight.

It is possible for enzyme treatment to take weeks, and even months, before results can be seen. Unfortunately, the patient has to bear the pain and discomfort of his or her condition while continuing therapy. Enzyme treatment demands great patience and perseverance from the beginning, but the end results make it all worth the wait.

Consistency in enzyme treatment is also of the utmost importance. Much like the treatment of MS, it is extremely important to be aware of progression from one phase of arthritis to another. See your doctor the moment you notice any change, such as a cold, influenza or fatigue. If your condition gets worse, you may need more enzymes. In the acute phase, more immune complexes are present in the joint fluid, cartilage and blood. A more aggressive enzyme treatment allows the enzymes to attack the immune complexes.

Get the Jump on Arthritis

In order to jump-start your day, and revitalize your life, arthritis patients should follow the Five-Step Jump-Start Plus Enzyme Program, which includes:

1. Cleansing the body (through fasting, improved elimination and detoxification), while avoiding internal and external pollutants, which build up toxins in the body
2. A rejuvenating diet
3. Enzymes and other supplements to restore and stimulate bodily functions and the immune system
4. Aerobic exercise, such as cycling, walking or running to the best of your ability
5. Positive mental attitude through mind and spirit power

Detoxification

The primary step in the Five-Step Jump-Start Plus Enzyme Program is detoxification of the body by ridding it of harmful toxins and encouraging regular bowel movements. The detoxification process is twofold: 1) aid the body in eliminating toxins (from the environment or arthritis itself), and 2) avoid environmental toxins as much as possible. Toxins originate from within when the body is unable to rid itself of waste products, when there is incomplete oxidation in the tissues or poor nutrition, and/or when some disease condition exists. Any disease puts heavy demands on the liver to get rid of waste. This is a huge problem when fighting arthritis or any other chronic disorders. (See the chapter on detoxification for further information.)

Diet

Many of those with arthritis have allergies to foods of the nightshade family. They have problems tolerating bell peppers, eggplant, paprika, potatoes, and/or tomatoes, which often cause their symptoms to increase.

The following are dietary guidelines and changes in lifestyle for the care of arthritis:

- Include foods such as whole grains, poultry and fish, which are rich in vitamin B3 and enzymes.
- Limit all red meats.
- Avoid pork and fatty meats, as well as smoked or barbecued foods.
- Lower your consumption of enzyme-inhibiting foods (soy bean products, peanuts, lentils, etc.).
- Eat fresh yogurt with active cultures daily.
- Load up on raw fruits and vegetables, whenever possible.
- Limit the intake of strong condiments and spices.
- Use garlic and onions if easily digested.
- Avoid common table salt. Use kelp and sea salt in moderation.
- Avoid refined sugar and refined carbohydrates such as pies, cakes, and desserts.
- Incorporate foods high in fiber, such as bran, whole grains, etc.
- Drink fresh raw juices, preferably vegetable (carrot, celery, parsley, beet, etc.), but avoid excessively cold or iced beverages.
- Eat small meals throughout the day, as opposed to three square meals of large proportion.
- Avoid food additives, preservatives, and food coloring.

Supplementation and the Immune System

It is important to use immune enhancers such as vitamins, minerals, homeopathic preparations and enzymes to build up your immune system. Don't forget antioxidants, such as zinc, selenium, vitamins A, C, and E, plus bioflavonoids.

Intensive Enzyme Therapy

Enzyme therapy is useful in the treatment of arthritis (see the book, *A New Look at Chronic Disorders and Systemic Enzyme Therapy*). Enzyme therapy should be a daily ongoing process, for a month, a year, a lifetime. Some individuals may have deficiencies in digestive enzymes and/or hydrochloric acid. Therefore, enhancers are needed in order to restore digestion to normal.

Exercise

Exercise forty minutes every day, three times per week and get plenty of natural sunlight (fifteen minutes daily). Avoid radiation from television, X rays or microwaves. This regimen should become a way of life to improve and maintain your health.

Positive Mental Attitude

As with anyone suffering from a chronic disorder, arthritis sufferers can sometimes get disheartened. Don't let this happen to you! A positive mental attitude can make all the difference.

In order to survive arthritis, you don't have to take gold or corticosteroids and suffer their immunosuppressive effects. Reduce inflammation, fight free radicals and boost your immune system with the help of enzyme therapy and the Five-Step Jump-Start Plus Enzyme Program.

Remember, enzymes *can* help diffuse:

1. Inflammatory rheumatic conditions
2. Soft tissue rheumatic disorders
3. Degenerative joint disease

References

1. CDC National Center for Chronic Disease Prevention and Health Promotion, Retrieved from http://www.cdc.gov.nccdphp/arthritis.htm, 26 June 1999.
2. C. Steffen, J. Smolen, et al., "Enzymtherapie im Vergleich mit Immunkomplexbestimmungen bei Chronischer Polyarthritis," *Zeitschrift fur Rheumatologie* 44:51 (1985).
3. I. Horger, V. Moro, W. van Schaik, "Zirkulierende Immunkomplexe bei Polyarthritis-Patienten," *Natur-und Ganzheitsmedzin* 117 (1988).
4. Wilhelm Glenk and Sven Neu, *Enzyme Die Bausteine des Lebens Wie Sie Wirken, Helfen und Heilen.* Munich, Germany: Wilhelm Heyne Verlag, 1990.
5. Klaus Uffelmann, "Enzyme treatment of soft tissue rheumatism," from lecture, First International Conference on Systemic Enzyme Therapy. Univ. of Klagenfurt, Klagenfurt, Austria. September 12, 1990.
6. F. Singer and H. Oberleitner, "Drug Therapy of Activated Arthrosis. On the Effectiveness of an Enzyme Mixture Versus Diclofenac," *Wiener Medizin Wochenschrift* 146(3):55–58 (1996).
7. F. Singer, "Aktivierte Arthrosen Knorpelschonend Behandeln," In: Medizinische Enzyme-Forschungsgesellschaft e.V. *Systemische Enzymtherapie, 10th Symposium*, Frankfurt, 1990.
8. Gert Klein, "Enzyme Treatment of Rheumatoid Arthritis," from lecture, First International Conference on Systemic Enzyme Therapy. Univ. of Klagenfurt, Klagenfurt, Austria. September 11, 1990.
9. Gert Klein, G. Pollmann, and W. Kullich, "Klinische Erfahrungen mit der Enzymtherapie bei Patienten mit chronischer Polyarthritis im Vergleich zur oralen Goldtherapie," (Clinical experience with enzyme therapy in patients with rheumatoid arthritis in comparison with oral gold) *Allgemein Medizin* 144–147: Oktober (1990).

CHAPTER 14

Oh, My Aching Back!

PAIN IN your neck, across your low back? Hurt to bend or twist? Sharp, stabbing pain in your back upon coughing or sneezing? Can't straighten up? These and others are symptoms of back pain.

Chronic back conditions include intervertebral disk disorders, curvature of the back or spine, and other chronic back impairments such as permanent stiffness or deformity of the back or repeated trouble with the back.[1] Chronic back conditions are both common and debilitating. The annual incidence of lower back pain is 5 to 14 percent of the American people, with the depressing note that 60 to 90 percent of Americans will experience chronic back pain some time in their life. It is estimated that lower back pain costs the nation $50 to $100 billion every year.[2]

Chronic back conditions rival arthritis and heart disease as a major cause of activity limitation causing loss of income, inability to work or be active, and a decline in self-worth. Therefore, a person with back pain is not a happy camper—nor is she alone.

World-famous back expert Dr. Rene Caillet states: "Of the numerous musculoskeletal disability conditions, the complaint of lower back pain is undoubtedly predominant."[3] "Estimates of hours lost to industry, hours of disability, and money paid for medical care and disability compensation are astronomic."

"Because our generation seems to have such a predisposition to cervical spine and lower back pain, the correct conservative management of these problems has never been more important than it is today," according to Frank E. Stinchfield, M.D., professor of orthopedic surgery.[4]

An initial attack of back pain is frequently precipitated by business strain, emotional problems, tension, unaccustomed sports or physical activities, or unaccustomed work. These causes may be combined with, or masked by, any true mechanical factors, but rarely are they completely missed. A minor trauma such as sneezing, turning, or picking up a briefcase may be inadequate to produce harm in "normal" backs. But any of these or making a sudden movement may be sufficient to cause the first attack of back pain. Further back attacks can recur many times without any known cause.

As time passes, the role of back muscles in homeostasis (and their importance in body balance) is becoming clearer every day. However, the effect of the endocrine, metabolic or cardiovascular systems, as well as emotions, on back muscles is not fully appreciated. Its role in a number of biomechanical deficiencies is also important. The importance of muscle function in back pain, poor posture, and other back problems has been stressed in the past by a number of researchers.

The average sedentary worker is subjected to a tremendous amount of stress. High pressure work at school or office, noise, telephones, traffic and competition in every phase of life all contribute to keeping us in a high state of stress. Our bodies, especially our muscles, are prepared to attack or run away, but seldom do we actually perform the tasks for which our body was constructed. We are forced by morals and the rules of society to "grin and bear it." That is, take whatever society has to give without striking back. This has a tremendous effect on our body as a whole, and specifically, on the muscles of our backs.

TENSION SYNDROME

We lead lives of caged animals in our civilized cities, with little opportunity to respond to outside irritations. Emotional problems are often an added burden and increase the need for release, which we must hold in. Since we do not have outlets through heavy exercise as our ancestors did, and since civilization does not approve of the natural response through "fight or flight," our muscles store up the tension. This constant tension deprives the muscles of elasticity and shortens them. Once the lack of physical activity has weakened the tense muscles, and muscle tightness has reached a sufficiently high level, the stage is set for the first episode of back pain. Even a minimal act of picking up a pencil or a piece of paper may trigger the first attack of back pain. This attack leaves the muscles stiffened and weakened and predisposes the back for the next painful attack, which in turn, will compound the symptoms. The vicious cycle has begun!

The spasm and muscle imbalance is increased and frequently there are local disturbances in the muscle itself once minor episodes of back pain have accumulated. Painful episodes increasing in both intensity and frequency result from the appearance of localized painful areas and tenderness of the constantly tightened muscle.

The patient's fear of disability, discomfort and anxieties can cause complications due to additional nervous tension. The stiff neck and low back syndrome have now reached the point where treatment is time-consuming, difficult and frequently hopeless.

At this point, the individual finds it much harder to return to former activities or even begin an active life. It becomes increasingly difficult to tolerate the muscular discomfort that accompanies any physical activity. This added hardship of retraining increases the problems of attitude as well as muscle, particularly in connective tissue.

Connective tissue is the largest component of the human body. Tissue is found throughout the body. Adipose tissue, cartilage, bone, collagen, and elastin make up the connective tissue system. Connective tissue is highly complex and specialized tissue which nourishes, supports, and defends against infection and trauma. Therefore, blood vessels and lymphatic vessels are essential for nutrition, repair and defense.

The warp speed with which our society is changing has caused back pain to be the "disease of civilization." We have become couch potatoes. We have failed in adapting to change in modern society. The couch potato syndrome is exemplified by our sedentary life in recent times and has caused a host of diseases classified as hypokinetic diseases or diseases of civilization.

The direct correlation between physical activity, calorie consumption and weight increase was shown by the late Jean Mayer. He felt overweight individuals have a minimal amount of activity in comparison to the amount of food (calories) eaten. This is associated with elevated insulin levels, elevated low-density lipoprotein cholesterol, predisposition to diabetes and atherosclerosis, and depression of enzymatic activity. A starvation diet is necessary to avoid weight increase below that minimum of activity. The relation between emotional stability on one hand, and sufficient activity and physical fitness on the other, has been noted in other studies.

Thus, lower back pain emerges less often as a localized mechanical problem than as a disease complex symptom.

However, mechanical back pain can be a real pain. There is no denying the orthopedic, neurological and X-ray findings of a protruded or ruptured disc and the like. Inflammation, heat and swelling are real causes of pain.

Therefore, back pain is a disease of civilization. It has many causes. Most back and neck pain is of muscular and other soft tissue origin.

However, back pain has endocrine, psychological, skeletal, occupational and environmental factors which may result in muscular problems. Therefore, to permanently overcome back pain, all aspects of lifestyle must be considered in order to achieve permanent recovery. This change in lifestyle should include the Five-Step Jump-Start Plus Enzyme Program.

Further research has shown that proteolytic enzyme use will: 1) speed up the inflammatory process and bring it to a conclusion; 2) help clean up the waste products in the area; 3) decrease pain and swelling; 4) dissolve any small blood clots floating nearby; 5) improve the supply of nutrients to the tissue, improving circulation; and 6) aid in easing blood flow.

Even the most fit of individuals can have a back injury. We do stupid things. We don't use the proper body mechanics in lifting. We reach out too far from the waist in picking up too heavy an object. We put undue stress on the back and bingo—back injury.

What happens? Stretching and/or tearing of the back muscle fibers (and possibly ligaments) may occur. The soft tissue cells and capillaries (smallest blood vessels) in the area tear and rupture. The whole process of inflammation takes place with pain, swelling, heat and possible discoloration (a red or purple color). Pain might be so severe we have difficulty moving (loss of function).

An example of this is Glen, a fifty-year-old in top physical condition. Glen is a world-famous cross-country coach from a large metropolitan area. A former Olympic coach, his high school cross country team has perennially won the national championship. Glen injured his back (attempting to pick up a safe!) and had suffered constant back pain for over three months. Although in constant agony, Glen ran five miles per day. He started taking enzymes. Three *days* later, he no longer experienced discomfort and could run without pain.

The following is a letter from a patient to his doctor regarding a "slipped disc," back pain, and enzyme treatment.

Dear Doctor:

I am writing this letter to you as a testimonial to the treatment that I received from you in 1980 when I had injured my low back while coaching soccer at — College. My injury occurred on Monday afternoon, November 16, while my team was practicing indoors. We were scrimmaging in the gymnasium and I was playing goalie. A shot was taken on goal and I blocked it with the inside of my right thigh. The force of the impact abducted and extended my thigh to the side. It also caused my lumbar spine to project forward putting stress on the fifth lumbar vertebrae, resulting in a spondylolisthesis of the fifth lumbar vertebrae. At that instant I was unable to straighten my leg at the knee or to lower my thigh, due to spasms in the calf muscles and hip flexor.

My wife picked me up and took me to a hospital emergency room, where I got pain medication and muscle relaxers. I was told to stay in bed for two weeks. By Thursday I did not have any reduction in symptoms and literally had to crawl to the lavatory. That is when I called you for advice. I saw you the next day (on Friday). The X rays showed a 12 millimeter slippage of the L5 vertebra forward, beyond the anterior lip of the sacrum. On the next Monday, I attended your clinic and the treatment was the same, with the addition of giving me a jar of proteolytic enzymes as I was leaving. Your instructions were to take a certain number of enzyme tablets daily. Since the Friday and Monday treatments, I did not have much relief from my pain, although there was some reduction. I took the pills on that Monday and when I woke up Tuesday morning I had literally lost almost all of the pain. I could stand erect, next to my bed, without having to bend over. The reduction of pain was remarkable.

As a researcher, a scientist, and Ph.D., I realize that studies would need to be done to form a definitive link between the cause and effect relationship of these proteolytic enzymes with the pain reduction, but as an N of one, I am convinced that the enzymes relieved my symptoms.

So you lifted a box out of the car trunk and hurt your back. It hurts to move, so you don't move. With reduction of activity comes a decrease in metabolic rate and in circulation. This can lead to an increase of toxic waste in our cells, circulatory system, and all other systems. There is a buildup of fibrin along the arterial walls, plaquing and a general overall stasis (traffic jam) in the cells and roadways to and from the cells. Nutrients can't get in and waste products can't get out. If this stasis occurs in the back muscles, inflammation, swelling, and pain result.

The predisposing physical problems can be complicating factors such as one leg shorter than the other, a hemi-vertebra (where the fifth lumbar vertebra is fused to the tailbone or sacrum on one side), having four or six lumbar vertebrae instead of five, being overweight with the lower back curved forward, arthritis, and/or muscle spasms.

Research shows that the most effective way of treating back sprains and strains is by using chiropractic or naturopathic care. The more complicated back conditions should employ a team approach (chiropractic, naturopathic, medical physician, etc.).

However, enzyme therapy can speed the course of lower-back rehabilitation and reduce the pain of many back problems. Enzyme therapy can be a definite aid in swiftly bringing the inflammatory process to an end, while reducing or eliminating the swelling and pain.

As we know, the process of inflammation is governed by numerous enzymes, a considerable number of which are the body's own proteolytic enzymes. They eliminate the inflammatory debris and initiate the restitution.

This process is supported and accelerated by administration of oral enzymes. They lead, in a biological way, to inhibition of a further spreading of the pathologic process and considerably reduce the duration of back pain. The thrombolytic effect of enzyme combinations is due to an activation of the body's own proteolytic system and to the fibrinolytic activity of the proteolytic enzymes.

The effects of proteolytic enzymes on the back are as follows:

Rapid depolymerization of the inflammatory debris
Dissolution of microthrombi

Restitution of micro-circulation
Increase of the tissue permeability
Reduction of edema

Thus,

Back pain stops instantly or is greatly reduced.
Inflammation is rapidly reduced.
The duration of the inflammatory process is considerably reduced.

Follow the Five-Step Jump-Start Plus Enzyme Program. Be sure to include detoxification, improved diet, nutritional supplements, exercise and stress-reduction through a positive mental attitude, as well as enzyme therapy. This program will revitalize your life and help you get a jump on back pain.

References

1. *Healthy People 2000.* Washington, D.C.: U.S. Department of Health and Human Services, Public Health Service, 1991.
2. Centers for Disease Control. Retrieved from http://www.cdc.gov/niosh/diseas.html.
3. Rene Cailliet, *Soft Tissue Pain and Disability.* Philadelphia: F. A. Davis Co., 1977.
4. Hans Kraus, *Clinical Treatment of Back and Neck Pain.* New York: McGraw Hill Book Co., 1970.

CHAPTER 15

Enzyme Cancer Fighters

CANCER IS the second leading cause of death in the United States and accounts for one out of every four deaths.[1] Cancer strikes more frequently with advancing age, but many cancer deaths are premature and could be prevented or, if detected and treated at early stages, cured.

Although there are many different kinds of cancer, a few major types account for more than half of all cancer-related illness and death. This list includes cancers of the lung and bronchus, colon and rectum, breast, and prostate.

The age-adjusted death rate from cancer has been steadily rising. However, most of the overall increase is due to a rise in lung cancer rates.

Nearly one million cases of skin cancer are diagnosed every year In fact, skin cancer is the most frequently diagnosed form of cancer. Melanoma—the most serious form of skin cancer—kills over 7,300 people every year.[1]

Lifestyle, environment, and genetic factors, individually or in combination, can increase your risk of developing cancer. It is estimated that diet contributes to 35 percent of all cancer deaths, and tobacco to 30 percent. In addition to these, other contributing factors include: reproductive/sexual behavior (7 percent), occupation (4 percent), alcohol (3 percent), geophysical factors (3 percent), pollution (2 percent), industrial products (1 percent) and medicines and medical procedures (1 percent).[1] Reducing fat intake and increasing consumption of fruits, vegetables, grains and dietary fiber (as well as reducing tobacco use) hold the greatest promise as strategies to reduce cancer incidence and ultimately, cancer mortality.

What role do enzymes play in cancer treatment? Whatever the causes of cancer, and however it is eventually understood, the study of enzymes will have a contribution to make.

The use of enzyme therapy on cancer probably began around the turn of the century. Dr. John Beard, a Scottish embryologist, was studying enzymes and their action in reproduction and growth. He knew the importance of enzymes in the growth process and reasoned that because cancer is simply uncontrolled cell growth, perhaps an enzyme deficiency was involved. He studied the use of a pancreatic extract in treating cancer and found that the extract could inhibit cancer cell growth. Since young animals require the most energy for growth, Beard believed they would have the most powerful enzymes. He took the pancreas of newborn lambs, hogs, and calves, minced them, rinsed the pulp, and filtered and pressed out the fluids containing concentrated enzymes.

Beard injected the fresh pancreatic juice (from the slaughtered animals) into the veins or gluteal muscles of his cancer patients. When possible, he injected the juices into the tumor itself. His theory was challenged by his colleagues. However, he did not allow himself to be discouraged. He knew the majority of his patients were in the terminal stages of cancer, and that any success, even the most remote, was promising. He noted the cancerous growth was inhibited and patients survived longer when enzymes were used.

From his findings on the treatment of 170 patients, he wrote the book, *The Enzyme Treatment of Cancer and Its Scientific Basis.* This manuscript has been the foundation for subsequent work in the area of enzyme therapy with chronic disorders.

Beginning in the 1950s, Dr. Max Wolf, with his collaborator, Helen Benitez, treated cancer cells with enzymes in order to discover which enzymes were most effective in degrading cancer cells (while leaving healthy cells untouched).

He found that combinations of enzymes seemed to work synergistically and were more effective in treating cancer, as opposed to the use of individual enzymes. He administered various enzymes to about 50,000 cancer patients over a period of twenty-five years and found the most effective to be a combination of beef pancreas, calf thymus, *Pisum sativum* (a variety of peas), *Lens esculenta* (a variety of lentils), papaya, and mannitol (a carbohydrate used as a carrier).

HOMEOSTASIS

How does cancer happen? The formation of cancer cells is due to many varying factors. There is interference in body cell production by certain chemical and physical mechanisms. Tiny cellular structure changes cause alterations which lead to new protein production. Cancer cells result from these malformed, abnormal cells. The cancer cell growth becomes uncontrolled, no longer subject to our body's defense systems.

On the positive side, these abnormal cells can be recognized as foreign because of their unusual structure. Our body's defenses can search out these cells, attacking, destroying, and eliminating them by enzymatic action.

Here comes the tightrope, the balance in life. At any given moment, our healthy body has thousands of cancer cells. But, this does not mean we have cancer. The poorly constructed degenerate cells are continually recognized by our body's defense systems, and either engulfed and eliminated by our predatory macrophages, or attacked by our antibodies. The cancer cells are disintegrated by a group of killer enzymes which answer the call for assistance.

Homeostasis is maintained. There is an equilibrium between cancer cells produced and cancer cells destroyed. As more new cancer cells are produced, about as many are recognized and destroyed. A few of these individual cancer cells survive, but they just float about aimlessly in the blood. They find nowhere to adhere to cell walls and then simply die.

Like an ominous, deadly smog hanging over us, this balance can be upset by negative environmental factors, such as pollution, aging, medications which weaken the immune system, nicotine and other street drugs, poor living and eating habits, or radiation. These factors make our bodies more susceptible to cancer by weakening our defense mechanisms. This can cause free radical formation and interferes with our "jump-starting" enzymes.

Dr. David J. Lin states in *Health News and Review* that a normal cell contains genes which tell it when to stop growing.[2] If these genes are mutated, such as by free-radical destruction, then the stop instruction is lost and the cell will continue to grow out of control. Think of a busy corner at rush hour. The stop light allows the traffic to safely pass through from divergent directions. Now think how chaotic it would be if someone removed the stop light so that cars would continue through the intersection without caution or concern. It would be disastrous. So is the action of free radicals in our bodies.

When our defenses are weakened, we can't hold the increased number of enemies in check. Because of this, more cancer cells remain free. They can adhere to cell walls, hide, and multiply, a particularly dangerous trick. Cancer cells realize the body's defenses can recognize and destroy them because they are different. They coat themselves with a thick, gluey, fibrin layer which hides their antigens and allows them to remain unrecognized, escaping detection.

These sticky cancer cells can attach themselves to any tiny bend in the vessel wall, coating themselves with more fibrin and multiplying under their camouflage. In this way, millions of cancer cells are formed, unexpectedly breaking through vessel walls and penetrating into the tissue, performing their deadly tasks.

The absence of certain enzymes is one reason why cancer cells can grow. Some enzymes are capable of stripping the fibrin from the cancer cells, laying bare their antigens and, thus, paving the way for their destruction by macrophages and the whole immune system. If you are deficient in certain enzymes, the cancer can grow. The presence of these enzymes holds the cancer cells in check. The more cancer cells produced in the body, the more enzymes the body needs.

Like wild children, cancer cells have no inhibitions. They are also wily and clever, using every trick in the book to outwit the immune system, our body's police force.

The cancer cells want to stop the body from alerting the immune system of their presence. They also want to keep enzymes from tearing away their protective fibrin coating. Their aim is to be the road block to enzymes and the immune system.

As stated before, immune complexes are formed when the cancer antigens and the body's antibodies join. Our ever-present PacMen (macrophages) gobble up and devour immune complexes, thus keeping their numbers in check. However, if our PacMen become overworked and outnumbered, they can't do their job, allowing undegraded immune complexes to remain in the blood and lymph.

TUMOR NECROSIS FACTOR

Some bacteria produce substances that stimulate macrophages, causing release of a factor that attacks cancer cells and destroys them. This concept was originally developed by Dr. William B. Coley, some one hundred years ago.

This substance, tumor necrosis factor (TNF), not only attacks cancer cells selectively, but is also active against cells that have been infected by viruses. One action of the immune system, when suffering from bacteria, fungi, or viral infections, is to release TNF.

Our immune system becomes depressed by our over-extended use of antibiotics, cortisone, and other infection-depressing drugs. This overuse effectively weakens our own natural method for destroying cancer cells and viruses.

Most doctors practicing natural medicine recommend that slight infections (which usually only last a few days) be allowed to take their normal course (under supervision) and be treated using natural remedies. Reducing the temperature of every minor fever interferes with the immune system's battle against the pathogen.

Our bodies can release small amounts of TNF without bacteria or viral stimulation. The body secretion of TNF can also be stimulated by some vegetable components in our diet including cabbage juice and ginseng. Even seaweed extracts can enhance production of tumor necrosis factor.[3,4]

Scientists have learned the best way to combat or protect against cancer is to keep the body in optimal condition, with natural methods. For this reason, millions of dollars are being spent in search of new, natural substances, discovering their basis, and ensuring their safety.

Enzymes

Many scientific articles have been published concerning the use of enzyme therapy for various forms of cancer. These articles provide the interested doctor and patient with valuable information.

For example, Dr. Lucia Desser (Austrian Cancer Research Institute, University of Vienna) studied the use of certain enzyme mixtures and their effects in stimulating macrophage TNF release in cancer patients.

Dr. Desser treated cells and found that enzyme mixtures and individual enzymes caused considerable TNF secretion.[5]

Since Dr. Desser's research, efforts have been made to discover exactly which types of cancer respond best to the enzymatically-stimulated TNF. Further, the TNF stimulation requires a differing enzyme dose depending on the type of cancer involved. For example, some cancers die completely after the administration of relatively low doses of enzyme mixtures, while other forms of cancer require much higher doses for longer periods of time.

Pancreatic cancer is one of the most serious forms of cancer, partly because it occurs without symptoms until it is in advanced stages. This form of cancer is also quite deadly. According to the American Cancer Society, the one-year relative survival rate is only 20 percent; the five-year rate is only 4 percent.[1] Studies on enzyme therapy in the treatment of pancreatic cancer are quite promising. Researchers Gonzalez and Isaacs conducted a two-year pilot study to access the effect of proteolytic enzymes in ten patients whose cancer was deemed to be inoperable. At one year, the survival rate was 81 percent; at two years, the survival rate was 45 percent. According to the researchers, four subjects have survived three years. The researchers believe large doses of pancreatic enzymes, along with an aggressive nutritional therapy, "led to significantly increased survival over what would normally be expected for patients with inoperable pancreatic adenocarcinoma."[6]

Dr. Ottokar Rokitansky recommends treatment of patients systemically with hydrolytic enzymes prior to surgery and injects some patients peri- and intratumorally.[7] It has been his experience that enzyme treatment induces a clearly-increased defense mechanism (in the body) against the tumor cells.

In addition, enzyme therapy has a definite place in prophylaxis, treatment, and especially in the follow-up care of cancer patients, according to Dr. Heinrich Wrba of Vienna, Austria.[8]

Rokitansky believes enzyme treatment is effective in conjunction with surgery and chemotherapy because:

1. Cancer cells are more sensitive to proteolytic enzymes than normal body cells.
2. Enzymes dissolve the fibrin coating on the cancer cells, allowing the body's defenses to function better.

3. Enzymes can diminish the stickiness of the cancer cell, preventing the formation of metastases.[9]

Although enzyme therapy seems promising, more research needs to be conducted in order to validate some of its specific applications.

In order to prevent possible recurrence of cancer, some doctors recommend measures to help strengthen the whole body. That is, the patient should incorporate a whole food diet (emphasizing fresh fruits and vegetables), vitamins, trace elements (especially zinc, copper, and selenium), detoxification, and the administration of herbal teas.

Enzyme therapy should be continued. Since the new cancer cells are constantly forming, they must be destroyed when they are still small metastases containing small numbers of cells and still searching for an anchoring place in the blood or lymph vessels. Metastatic prophylaxis should always include enzymes.

A healthier lifestyle should be continued, thus strengthening the body's defenses. With a preventative program, the patient discharged from the hospital as cancer free and the at-risk cancer patient have a good chance of keeping the disorder under control.

Breast Cancer Surgery

We should briefly discuss a disorder associated with breast cancer surgery and its aftercare, *lymphedema*. Many women who have undergone breast removal surgery (and radiation therapy to the armpit) subsequently suffer a lymphedema of the arm. This swelling can sometimes reach massive proportions and has considerable effect on the patient's quality of life. A lymphedema can also interfere with healing.

Drs. Glenk and Neu feel it is possible to prevent lymphedema development with some degree of certainty if enzyme mixtures continue to be taken.[10] Anyone who has experienced the suffering that such a lymphedema can cause realizes the importance of long-term prevention.

Another application of oral enzyme therapy is just as important. The Janker Radiation Clinic in Bonn, Germany investigated many women (especially younger women), who felt a lump in their breast and suspected breast cancer. But lumps and painful inflammatory changes in the breast are no sure sign of malignancy. To be on the safe side, tissue samples are often taken or operations performed when the tissue is actually a benign growth.

The Janker Clinic's test results indicated that the benign lumps and painful processes disappeared in 90 percent of the women using enzymes with vitamin E. Should the lump later reappear, a repeat course of enzyme tablets and vitamin E usually had the same positive result. This seems to be an almost harmless recipe, but one which is certainly capable of freeing many women of anxiety, thus sparing them other, more unpleasant, treatments.[10]

Cancer Fighting in Review

Enzyme therapy improves the body's defense powers and retards the growth of cancer cells.

Enzyme mixtures attack two dangerous properties of the cancer cells. First, they can degrade the fibrin, unmasking the cancer cell antigens and leaving them open for attack by our immune fighters. Secondly, they can remove the "glue" with which cancer cells attach themselves to vessel walls and tissues. The enzymes can break up immune complexes and chew up fibrin.

This means that people with an increased risk of cancer can reduce that risk by taking enzymes. Further, it seems obvious that certain enzymes should be taken before and after a cancer operation. This will compensate for the weakened immune system which always follows every surgical intervention.

Enzyme therapy, during the treatment of cancer, is gaining importance in hospitals. Numerous oncologists in Germany, Italy, France, and the United States know the value of these enzymes.

If radiation and/or chemotherapy are necessary, lower doses than usual can be just as effective with concurrent enzyme administration. In addition, a higher enzyme level protects the body from the feared radiation hangover and from some of the other unpleasant radiation and chemotherapeutic side effects.

Whenever operations, radiation, or chemotherapy are impossible (or no longer of value), enzyme therapy can at least have a pain-alleviating effect. In addition to tablets, enzyme mixtures can be injected directly into almost all tumors (accessible to the needle). This can initiate complete tumor disintegration. The administration of an enzyme mixture is often accompanied by an improvement in the morale of cancer patients. They regain their appetite, put on weight, cease to be depressed, and feel considerably more alive both physically and mentally.

Enzymes substantially reduce the occurrence of metastases and the recurrence of cancer. They make malignant tissues more sensitive to radiation, and at the same time, reduce pain and other side effects of radiation. When used in conjunction with surgery, they reduce postoperative pain and hemorrhage. In terminal patients, they ease the pain of the last days, quite often eliminating the need for strong drugs, which can make the patient somewhat less than human. They are useful in conjunction with many other forms of therapy. They are effective against a broad spectrum of malignant diseases, including cancers of the reproductive organs, the breast, the skin, the digestive system, connective tissues, and others (including Hodgkin's disease and leukemia).

References

1. *Cancer Facts and Figures 1999: Basic Cancer Facts.* Atlanta: American Cancer Society, 1999.
2. David J. Lin, "Antioxidant Nutrients May Act as Armor Against Free-Radical Damage, Cancer," *Health News and Review*, 3(3):22 (1993).
3. W. Komatsu, K. Yagasaki, Y. Miura, and R. Funabiki, "Stimulation of Tumor Necrosis Factor and Interleukin-1 Productivity by the Oral Administration of Cabbage Juice to Rats," *Bioscience, Biotechnology, and Biochemistry* 61(11):1937–1938 (November1997); T. P. Smolina, T. G. Orlova, O. N. Shcheglovitova, and N. N. Besednova, "Interferon-Inducing Action of Polysaccharide-Containing Biopolymers from Ginseng Root and Cell Culture," *Antibiotikii Khimioterapiia* 43(11):21–23 (1998).
4. J. N. Liu, Y. Yoshida, M. Q. Wang, Y. Okai, and U. Yamashita, "B Cell Stimulating Activity of Seaweed Extracts," *International Journal of Immunopharmacology* 19(3):135–142 (March 1997).
5. Lucia Desser and Alexander Rehberger, "Induction of Tumor Necrosis Factor in Human Peripheral-Blood Mononuclear Cells by Proteolytic Enzymes," *Oncology* 47(6): 475–477 (1990).

6. N. J. Gonzalez and L. L. Isaacs, "Evaluation of Pancreatic Proteolytic Enzyme Treatment of Adenocarcinoma of the Pancreas, with Nutrition and Detoxification Support," *Nutrition and Cancer* 33(2):117–124 (1999).
7. Ottokar Rokitansky, "Adjuvant Enzyme Treatment Before and After Breast Cancer Surgery," *Dr. Med* 1(2):16 (1980).
8. Heinrich Wrba, "New Approaches in Treatment of Cancer with Enzymes," from lecture, First International Conference on Systemic Enzyme Therapy. September 12, 1990.
9. Ottokar Rokitansky, "Adjuvant Enzyme Treatment Before and After Breast Cancer Surgery," *Dr. Med* 1(2):16 (1980).
10. Wilhelm Glenk and Sven Neu, *Enzyme Die Bausteine des Lebens Wie Sie Wirken, Helfen und Heilen.* Munich, Germany: Wilhelm Heyne Verlag, 1990.

CHAPTER 16

For Women Only

FOR TOO long, we have ignored the illnesses of women or at least have put their illnesses on the back burner. The increasing investigation of female problems of the breasts and gynecological considerations is a step in the right direction.

German gynecologists are now using enzyme mixtures to treat certain female diseases which until recently, had limited success with traditional treatment. Doctors all over Europe today are using enzyme mixtures, either in combination with these traditional treatments, or by themselves. This is exciting.

GYNECOLOGICAL APPLICATIONS OF ENZYMES

Professor Doctor Friedrich-Wilhelm Dittmar[1] examined systemic enzyme therapy and its applications to gynecology. Diseases of the pelvis, mastopathy, malignant neoplasms, and endometriosis genitalis externa were among the selected gynecological complaints examined. Dittmar specifically investigated PID (pelvic inflammatory disease), in a double-blind, randomized study of 100 patients. After fourteen days of enzyme treatment, the majority of patients were free of complaints.

Dr. Dittmar also reported on the use of enzymes in *adnexitis* (inflammation of uterine appendages), the ovaries, uterine tubes and ligaments. Adnexitis is an extremely serious condition which can potentially lead to infertility and sterility in young women. The treatment normally prescribed for acute adnexitis is antibiotics to support the healing process. Certain enzyme mixtures actually change the outer tubal area, and positively influence endothelial synthesis. When enzyme therapy is properly employed, surgically-corrected conditions remain stable and residuals are prevented.

Another application for enzyme therapy was discovered by Hunter and Simmons[2] in the treatment of *dysmenorrhea* (a common condition in women associated with menses). They both reported relief among most of the cases they treated.

Dr. Dittmar (a senior physician in the Department of Obstetrics and Gynecology at the District Hospital in Starnberg, Germany) has successfully used enzyme mixtures for a variety of female disorders. For example, benign alterations in the female breast (mastopathy), inflammation of the fallopian tubes and/or ovaries (adnexitis), diseases of the pelvis, malignant neoplasms, and endometriosis genitalis externa, are but a few of many disorders.

He has also treated endometriosis, a difficult condition associated with pelvic discomfort and painful menses. It is estimated that endometriosis affects 5.5 million women and girls in North America.[3] It is caused by benign deposits of cells originating in the lining of the uterus. These deposits grow beyond the uterus and become inflamed during hormonal cycles.

Dr. Dittmar has noted the time length and difficulty of using traditional treatments with acute, subacute, and chronic adnexal illnesses. He noted that patients suffering from pain are generally less active and less able to handle daily activities. Some of the problems include: adhesions, scarring, disturbed menstrual cycles, uncomfortable vaginal flora (discharge), degenerative symptoms in the affected ovaries, and the danger of sterility. Since their side effects are quite a problem for women of child-bearing age, the available drugs can only be given for short periods of time and with great reservations, according to Dr. Dittmar.

Because of these concerns, Dr. Dittmar searched for an alternative approach which would be acceptable to patient and doctor alike. Enzyme therapy seemed to be the solution.

With this in mind, Dr. Dittmar performed a double-blind, clinical study on 56 adnexitis patients. Results indicated a complete improvement for the enzyme patients and no change for those not receiving enzyme therapy.

Clinically, Dr. Dittmar uses enzyme mixtures as standard therapy in his hospital. He feels the most effective approach is to immediately give the patients anti-inflammatory enzyme mixtures on the first office visit, as an acute therapy and, in many cases, to avoid such late symptoms as sterility and chronic adhesive inflammations. During the acute phase of this condition, enzymes are considered adjunctive therapy, taken in association with antibiotics.

One of the major problems after childbirth is the pain (and scarring) from the episiotomy (surgical incision). Researchers have found proteolytic enzymes helpful to decrease the pain, as well as the scar tissue from this type of surgery.[4, 5]

Fibrocystic or Cystic Breast Disease

Fibrocystic disease occurs in almost 50 percent of premenopausal women.[6] A benign condition, it is the most common disorder of the breast. Because new cysts usually don't appear after menopause, it is believed that hormones may be involved in their appearance. Though the patient may not have any pain, sometimes pain or premenstrual breast discomfort accompanies this condition. Usually the cysts are found during breast self-examination.

Enzyme therapy has been used successfully with mastopathy (cystic disease of the breast), according to Dr. Dittmar. Mastopathy is reported to affect about half of all women of child-bearing age and causes benign changes in the tissue of the breast. Complaints include stabbing or dragging pain in the axillary region (armpit) as well as swelling, pain on tension, and tenderness with pressure. During menstrual periods, the complaints are more severe.

Some authors consider mastopathia fibrosa cystica to be a precancerous disease and previously, no sufficiently safe therapy had been available, according to Dr. Scheef of the Janker Clinic in Bonn, Germany.

Since positive results had frequently been obtained using proteolytic enzymes in treating artificially caused carcinomas and fibromas of the breast, Drs. Wolfgang Scheef and Konig studied 247 women over a period of five years.[7] They used three groups, one with enzymes only, one with enzymes and vitamin E, and one receiving neither enzymes nor vitamin E. Subjective findings including sensation of pain and tenderness plus sonographic examinations were used to evaluate results.

After six weeks of treatment, 65 percent of the women in the enzyme group and 85 percent of those taking a combined enzyme-vitamin E regimen were free of complaints. After one to six months, recurrences took place in some patients, but they quickly responded to renewed therapy.

Because of the tendency for recurrences, Dr. Scheef recommends enzyme maintenance therapy. Since there are no serious side effects, he feels continued enzyme therapy has a clear advantage over the prospects of breast cancer. What do you think?

Dr. Scheef has continued to use enzymes to treat women with mastopathy. He reports relief of complaints, thus reducing the anxiety and apprehension of possible breast cancer.

To reaffirm the findings of Dr. Scheef, Dr. Dittmar performed a double-blind study on 96 patients suffering from mastopathy.[1] Forty-eight patients were treated using an enzyme mixture and 48 with a placebo. After six weeks, the enzyme group had improved appreciably, whereas the complaints of the women in the placebo group remained unchanged. There was a reduction in the size of cysts in the enzyme group, although the number and size of benign cists in the placebo group had not changed.

Hormones are often administered to women suffering from mastopathy in order to alter the hormones of the body that may be encouraging cancer growth. The effect of one such hormone, Lynestrenol, was compared with an enzyme preparation in twenty-nine women with mastopathy. After two months of therapy, all (100 percent) of the women in the enzyme group had a significantly greater decrease in number of hardenings of the mammary gland, while only 78.6 percent of the Lynestrenol group experienced a similar decrease. The researchers believe that enzyme therapy "is an alternative, low-risk therapy for the management of mastopathy, which does not interfere with the already upset hormonal balance of the patients."[8]

With certain surgical techniques in the past, many cancer "nests" may have been accidentally seeded or left behind in the operative region with resulting risk of spread. Further, after radical mastectomy, a high percentage of patients were permanently affected both psychologically and cosmetically.

Viennese oncologic surgeon Dr. Ottokar Rokitansky conducted a ten-year study on 305 postoperative breast cancer patients, including a ten-year follow-up.[9] The results were impressive. He reported that the ten-year survival rate for patients with Stage I breast cancer was 85 percent. Remember, five years of cancer survival is referred to as the Five-Year Cure. Patients having more severe conditions with metastasis to the lymph nodes (Stage II cancer) reported a 75-percent survival rate. For the true magnitude of this study's success, please compare survival rates in other studies, using other treatment programs.

Another feared complication following breast surgery is lymphedema of the axillary region. Lymphedema is the accumulation of excessive lymph fluid and swelling of the tissue due to obstruction, destruction, or hypoplasia of lymph vessels. It is not the same as metastatic involvement of the lymph nodes or ducts, but of the lymphatic vessels, themselves. This condition can occur as a result of tissue damaged during breast cancer surgery. Plasma protein escapes into the surrounding tissue. As a result, there is swelling, pain, and inflammation. The swelling in the armpit causes increasing discomfort and limited arm mobility. The swelling can become as hard as a board and so increased in size that it compares in magnitude to elephantitis. A superimposed danger with lymphedema is another type of cancer relating to damaged blood and lymphatic vessels, called Stewart-Treves Syndrome, or lymphangiosarcoma.

BREAST CANCER

This is a woman's greatest cancer fear. Breast cancer is the most common cancer among women and is the second leading cause of cancer death in women. It is estimated that 175,000 new cases of breast cancer were diagnosed in 1999.[10] But women aren't the only ones who get this disease, as about 1,000 new cases will be diagnosed in men.[7]

Symptoms of breast cancer include any breast changes that persist (such as a lump, swelling, or any changes in the nipple).

The American Cancer Society recommends that women age forty and older have an annual mammogram and an annual clinical breast exam performed by a health-care professional.

The good news is that the five-year survival rate for localized breast cancer has risen from 78 percent (in the 1940s) to 97 percent today.[10] With advances in operative techniques, cosmetic breast reconstruction results, and improvements in chemotherapeutic and radiation methods, the chances of overcoming this condition have greatly improved.

Through effective surgical sealing of the blood vessels going to and coming from the postsurgical region, the major risks of metastasis are reduced. Death resulting from cancer cells spreading through the body by way of lymphatic and blood vessels is far greater than death from the tumor itself. In cases of more advanced or dangerous tumors (except in pregnant patients) Dr. Ottokar Rokitansky uses very high doses of vitamin A in combination with enzyme therapy.

It is normal to have concerns over breast cancer. However, preventative, conservative care (including enzyme therapy) might prove to be the light at the end of the tunnel.

Among other things, our goals in fighting cancer are to build up our immune system and to reduce the inflammatory process. A combined Five-Step Jump-Start Plus Enzyme Program involves the following:

1. Detoxify (detoxification, fasting, juicing).
2. Eat a well-balanced, jump-start diet.
3. Use enzyme, vitamin, and mineral supplements, to jump-start your day.
4. Exercise daily.
5. Have a positive mental attitude.

Please note, this program should be followed only after consultation with a well-trained physician.

References

1. Friedrich-Wilhelm Dittmar, "Enzymtherapie in der Gynakologi (Enzyme Treatment in Gynecology)," *Allgemein Medizin* 19:1568–1569 (1990).
2. L. B. Sheila and M. B. Glasg, "Bromelain and the Cervix Uteri," *Lancet* 2:1420–1422 (1960).

3. Endometriosis Association International, Retrieved from http://www.endometriosisassn.org/endoassn_main.htm.
4. S. E. Soule, H. C. Wasserman, R. Burstein, "Oral Proteolytic Enzyme Therapy (Chymoral) in Episiotomy Patients," *American Journal of Obstetrics and Gynecology* 95:820–823 (1966).
5. H. D. Bumgardner, and G. I. Zatuchni, "Prevention of Episiotomy Pain with Oral Chymotrypsin," *American Journal of Obstetrics and Gynecology*, 92(4):514–517 (1965).
6. Robert Berkow, ed., *The Merck Manual*. Rahway, N.J.: Merck Sharp & Dohme Reasearch Laboratories, 1987, 1718–1719.
7. W. Konig, "Erfarhrungen der Robert-Janker-Klinik, Bonn, Mit Systemischer Enzymtherapie und Emulgierten Vitaminen," *Acta Medica Empirica*, 37:11 (1988).
8. E. Rammer and F. Friedrich, "Enzyme Therapy in Treatment of Mastopathy. A Randomized Double-Blind Clinical Study," *Wiener Klinische Wochenschrift* 108(6):180–183 (1996).
9. Ottokar Rokitansky, "Ozontherapie und Enzyme bei der Chronisch-Arteriellen Verschlusskrankeit Systemische Enzymtherapie am 31.10.90," *Natur-und Ganzheitsmedzin*, Supplement:14 (1991).
10. *Cancer Facts and Figures 1999: Selected Cancer*. Atlanta: American Cancer Society, 1999.

CHAPTER 17

Of Interest to Men

JUST AS there are certain diseases that only occur in women, so are there diseases that occur only in men. A good example is those disorders that affect the prostate—disorders than can be helped with enzyme therapy.

The male prostate gland is about the size of a walnut. Its job is to produce a thick fluid that assists in the transport of sperm. If the prostate becomes enlarged, it can interfere with urination. This condition—called *benign prostatic hyperplasia*—affects nearly 90 percent of all men over 80 and half of those over fifty.[1] Treatment for benign prostatic hyperplasia varies from watching and waiting to drug therapy, surgery, or compressing or removing the prostate. The prostate can also become inflamed, leading to a painful condition called *prostatitis*.

PROSTATITIS

Prostatitis refers to four different disorders of the prostate: *acute bacterial prostatitis* (caused by a bacterial infection that is easily treated with antibiotics), *chronic bacterial prostatitis* (a condition in which bacterial prostatitis persists because of an underlying defect in the prostate), *nonbacterial prostatitis* (the most common form of prostatitis and the least understood, in which infection-fighting cells are often found in the semen and other fluids), and *prostatodynia* (similar to nonbacterial prostatitis, but without objective findings).

Standard medical therapy for prostatitis varies depending on the type. However, most therapy includes antibiotics. Russian studies indicate that an enzyme combination (containing papain, pancreatin, bromelain, trypsin, chymotrypsin, and the bioflavonoid, rutin) is effective in the treatment of prostatitis.[2] Enzymes are also well-known for their ability to improve antibiotic absorption and efficacy, when the use of antibiotics is advisable.

PROSTATE CANCER

Prostate cancer is the most common type of cancer in men. Nearly 180,000 new cases will have been diagnosed in 1999. The incidence of prostate cancer increases with age, with most cases (75 percent) occuring in men over the age of sixty-five. However, it can occur at an earlier age, so the American Cancer Society recommends that men over fifty have a digital rectal exam and a prostate-specific antigen (PSA) blood test every year. Individuals who have a close family member with the disease and those of African-American descent are considered to be at high risk and should have the tests earlier than age fifty.[3] Fortunately, the five-year survival rate is quite high (100 percent) if prostate cancer is diagnosed before it has a chance to spread. In fact, the survival rate for all stages of the disease has increased over the past twenty years.[3]

The PSA test is widely used for the detection and monitoring of prostate cancer. If prostate-specific antigens are detected in the blood at elevated levels, it is an indication that cancer may be present. A healthy PSA reading is typically 4 or less; a reading between 4 and 10 might be an indication that further tests are needed; while a level above 50 would indicate that cancer has metastasized (spread) throughout the body.[4]

Standard medical treatments for prostate cancer include surgery, radiation, hormone treatment, and chemotherapy. Radical prostatectomy is a surgical treatment in which the prostate is completely removed.

However, enzyme therapy can also help fight cancer. It does this by boosting the immune system's natural ability to attack cancer. Enzymes can also help unmask the cancer cells' antigens and remove the "glue" that cancer cells use to attach themselves to tissues. Enzymes are also beneficial when taken just before and after any cancer operation, because they help stimulate the immune system, reduce inflammation, and speed healing. When administered concurrently with chemotherapy or radiation, enzymes can help diminish the radiation hangover as well as some of the unpleasant side effects of chemotherapy.

References

1. Food and Drug Association. Retrieved from http://www.fda.gov/fdac/features/1998/598_pros.html.
2 V. Martynenko, "Wobenzym in the Combined Pathogenetic Therapy of Chronic Urethrogenic Prostatitis," *Lik. Sprava* 6:118–120 (August 1998); A. R. Guskov, I. D. Bogacheva, and G. B. Iatsevich, "Systemic (Vobenzyme Preparation) and Local Enzyme Therapy in Transurethral Drainage of the Prostate in Patients with Obstructive Forms of Chronic Prostatitis," *Urol. Nefrol.* 6:37–42 (November/December 1998).
3. *Cancer Facts and Figures 1999: Selected Cancers*. Atlanta: American Cancer Society, 1999.
4. U.S. Food and Drug Adminstration. Retrieved from http://www.fda.gov/fdac/features/1998/598_pros.html.

CHAPTER 18

Circulation Rejuvenator

> The heart in itself is not the beginning of life; but it is a vessel formed of thick muscle, vivified and nourished by the artery and vein as are the other muscles.
>
> *The Notebooks of Leonardo*, LEONARDO DA VINCI (1452–1519)

NEARLY 900,000 Americans die every year from circulatory disorders, which include heart diseases and cerebrovascular diseases. This is more than all the deaths from cancer, accidents, pneumonia, and influenza combined.[1] But circulatory diseases are not limited to the elderly. In fact, almost 140,000 Americans under the age of sixty-five die every year from circulatory disease.

Besides killing nearly as many Americans as all other diseases combined, circulatory disease is also among the leading causes of disability.[2]

It has been proved that high blood cholesterol and coronary heart disease are directly related. In one study, a drop of 9 percent in blood cholesterol levels equated to a 19 percent drop in the incidence of coronary heart disease![2] About sixty million American adults have high blood cholesterol levels requiring medical advice and intervention using diet as the primary treatment.[2] Enzymes can also help.

CIRCULATION—OUR BODY'S LIFE LINE

One of the most fantastic aspects of that wondrous machine we call our body is the circulatory system. Like the body itself, the circulatory system uses a series of checks and balances. When this control system is interrupted, disorders such as high blood pressure, high blood cholesterol, or hardening of the arteries can result. Tumors can form, embolisms can develop, or plaquing and narrowing of the arteries can occur.

These circulatory disorders, particularly acute thromboses and embolisms (known collectively as *thromboembolic vascular diseases*), are still the largest single cause of disease and death in the middle-aged and elderly populations throughout the Western world.

Four to six quarts of blood constantly flow through your blood vessels and are pumped by the heart through the arteries. The arteries gradually narrow into a vast network of capillaries, supplying every cell of your body. Some capillaries are so minute that the blood must be squeezed through the cells. After the blood passes through the capillaries, its return trip takes it through larger and larger veins fitted with nonreturn cupped valves. The route back to the heart and lungs is not easy, as the blood must course uphill, without much pumping assistance.

There are possibly millions of miles of blood and lymphatic vessels, and we can only estimate the total length of these vast interconnecting networks. The size of the vessels in your circulatory system varies from as fine as a microscopic thread to as large as your thumb.

Circulation is an ingenious, extremely complicated system of pumping and sucking valves and conduits. When in balance, the blood flows and the system functions without problems.

There must be a continual ebb and flow in circulation, just like everything else in life. There is a never-ending interaction between blood liquefying and blood solidifying. In order to transport the nutrients and metabolic wastes, and to nourish all the body's cells, the blood's equilibrium has to be maintained in the correct state of fluidity.

The interaction between liquefaction and coagulation in the blood vessels is one of the most important equilibrium systems in the body. There is a constant dynamic balance between dissolving of blood clots and clotting (coagulation of the blood). Clots (called thromboses) are formed in the blood vessels if fibrinolysis (fibrin breakup) is weakened. Circulating fragments of clots can cause a sudden blockage of the blood vessel (called an embolism). A tendency toward excessive bleeding results if fibrinolysis is increased. Thus, the maintenance of homeostasis by the circulatory system is extremely important. In fact, embolisms and thromboses can be life-threatening.

It would be impossible to live if your blood were too liquid. A cut finger, bleeding gums, a scraped knee, the smallest of injuries to this great network of vessels could cause our blood, our fluid of life, to pour out at the site of injury. We could bleed to death. Blood clots to counter this possibility. This clotting saves us from death. The blood becomes increasingly thicker and sticky at the injury site and ultimately coagulates.

The body uses a series of inhibitors and activators to control the equilibrium between clot dissolution and formation. In forming a clot, fibrinogen (the plasma protein) is changed to insoluble fibrin by the enzyme thrombin, and fibrin forms the clot. The proenzyme plasminogen exists in the blood but must be changed to plasmin (an enzyme) to dissolve the clot.

Fibrin Formation—The Sticky Stuff

Fibrin, the glue our blood produces for coagulation, is required by our body for more than just sealing wounds. A continuously thin layer of fibrin also coats the delicate internal blood vessel walls, protecting them from damage. Any small uneven vessel walls are smoothed by fibrin, allowing the blood to flow without creating any disturbances.

To accomplish this, the body produces about two grams of fibrin each day. Without a system of checks and balances, the blood vessel walls could gradually become coated with a thickening layer of this adhesive fibrin. It would become increasingly difficult for blood to flow through the vessels. Finally, blood flow would stop and stasis would follow. Luckily, we not only have a system of adhesive fibrin formation, but also fibrin dissolution (called fibrinolysis). This system's purpose is to maintain an equilibrium in blood flow by degrading excessive fibrin formation.

Blood becomes too thin if fibrinolysis increases too much. If this happens, you could bleed to death. One inherited condition of excessive bleeding is called hemophilia. On the other hand, too much fibrin production causes the blood to become too thick and sticky. Blood stickiness is a result of insufficient breakdown rate, or fibrin overproduction, and is the most frequent deadly companion of vascular disorders and fatal heart disease.

In an emergency, the body does not have sufficient time to produce fibrin from scratch (for example, to seal off a wound). In addition, it is important that the adhesive substance not be deposited in the wrong places. This could stop the flow of blood.

For the sake of safety, the fibrin-producing enzyme, thrombin, is not present in the blood in the active form, but (just like fibrinogen), is found in a safe form, called *prothrombin.* Should the alarm of injury sound, a series of reactions (catalyzed by enzymes) is immediately triggered. The final result is an activator for converting the sleeping prothrombin into *thrombin.* The activator functions rather like the second arm in a pair of tweezers. One-armed tweezers are not good for anything. But, once the second arm is present, the tweezers are effective. Thus, it is the same in the thrombin function.

Plasminogen is a harmless, inactive proenzyme with its safety catch on. To become active, it is converted to plasmin. For example, inactive plasminogen hears the wake up call and changes to the active enzyme, plasmin.

Plasmin chops off individual members of the giant fibrin protein molecule. This action causes the fibrin to break up into separate stable protein chains which can be transported away. The same process can serve different ends. Thrombin nips off a portion of a soluble protein molecule and an adhesive fibrin results. Plasmin nips off a bit of the adhesive fibrin and it dissolves.

This is a good example of nature choosing the best technology. However, nature is not immune to error. The particular danger is that fibrin dissolution may not take place because of an error in plasmin activation. Inadequate fibrinolysis can also cause stoppage of the lacrimal (tear) ducts, salivary and mammary glands. The tubes carrying urine will not open. Fibrinolysis also plays an important role in a woman's menstrual bleeding. If fibrinolysis failed, the liver, kidneys, and lungs would become jellified, would harden, and finally all organs would simply cease to function.

A further complication is that fibrinolysis takes place at different rates in different parts of our bodies. Thus, the fibrin-dissolving activity in the veins is greater than in the arteries, it is greater in the veins of the arms than in the legs. This is a result of plasmin level differences, that is, the amount and quality of plasmin to be found in the different regions (plasmin alone is capable of degrading the sticky fibrin into its components).

In general, plasmin levels decrease with increasing age. At sixty, we only possess a fraction of the plasmin that we had when we were young. The results are obvious. The blood flows more sluggishly, toxic debris tends to remain in the vessels, the vessels become narrower and harden. Injuries and other disturbances occur which tend to compound the problem until they cause disease, resulting in degenerative symptoms (such as insufficient blood flow to the heart, brain, or legs [varicose veins] etc.) that are virtually accepted as normal in older people.

Once a clot is dissolved, there is always a danger of clots reforming. The tissue in the areas where a clot dissolved remains damaged. Therefore, it is always a strong likelihood of new clot formation at the spot of an old clot.

As we walk the tightrope of life, we must learn ways to restore the natural equilibrium through the use of enzymes when it is lost.

Cholesterol—The Fatty Vessel Lining

The most important factor in arterial diseases is the depositing of cholesterol and other fatty materials on the lining of blood vessel walls. When this pathological condition develops, it is called *atherosclerosis*.

The linings of all arteries, from the aorta to the arterioles, are covered by a thin fibrin film. During fibrin synthesis, an equilibrium exists between the constant deposition of fibrin and simultaneous fibrinolysis (destruction of fibrin). While this equilibrium exists, the arterial circulation remains stable. When this equilibrium is disturbed, there is an excessive amount of fibrin and lipids. Swelling of the blood vessel walls and further deposition of lipids and fibrin follow. The opening of the blood vessels becomes narrowed, causing reduced blood supply. This disturbance causes many diseases, such as intermittent claudication and angina pectoris, and has a serious effect on the circulation of smokers.

Cholesterol stands accused as the principal criminal in all cardiovascular diseases, but it is a vitally important component of cell membranes and is a beginning material for many hormones. It is a fatty substance our bodies produce and which we consume in our daily food.

Cholesterol level is important. A cholesterol deficiency can injure one's health. But, the concentration of cholesterol in the blood is usually too high. An orally-ingested enzyme mixture makes it possible to lower the blood's cholesterol concentration as well as triglycerides (the body's biggest energy store particularly in fatty tissue).

Cholesterol takes a form that is not normally considered "fatty"—pointed crystals. When the blood contains many of these pointed cholesterol crystals, the blood pressure is high. The crystals press into the vessel walls, the vessel loses its elasticity, and becomes covered with the deposit. This is the beginning of arteriosclerosis (hardening of the arteries).

Another danger is caused by the pointed cholesterol crystals. The damaged vessel walls change their structure and are mistakenly identified by the immune system as not being part of the body. Rather the damaged vessel walls are seen as foreign material which must be destroyed and removed. An antigenic effect is possessed by the vessel walls and the body's defenses mobilize against it. Thus, an autoaggressive disease is created where the body fights against itself.

Hoodwinking the immune system and inflammation are always involved in an autoaggressive disease. There are several immunologically-caused vascular disorders. For example, chronic rheumatoid arthritis or lupus erythematosus can spread through the entire vascular system causing vessel wall inflammation. This leads to a condition called *vasculitis*. Enzyme mixtures have proven to be an excellent treatment for vasculitis.

Certain enzymes seem to lower cholesterol levels. A nurse who was closely monitoring her cholesterol level started taking certain enzymes for her lower back pain (and possible lumbosacral disc involvement). Not only did the lower back pain go away, but also the tested blood cholesterol and triglyceride levels decreased.

Enzyme Therapy

When the blood is too sticky, small clots form which are then deposited at any narrowing of the blood vessel (forming a barrier where each subsequent blood clot is deposited), until finally, blood can no longer flow through the vessel.

Plasmin activation (to dissolve fibrin) is thus a critical measure in combating the danger of venous and arterial blood flow disturbances. One of the most effective ways of promoting the dissolution of fibrin is undoubtedly the administration of fibrinolytic enzymes. These enzyme mixtures contain the faithful servants required for plasmin activation.

The benefits of enzyme therapy have been demonstrated in countless double-blind studies. Such enzyme mixtures are able to dissolve the little clots (called *microthrombi*) and normalize the blood flow equilibrium. Certain enzymes can exert a positive effect on the tendency of blood platelets to clump and harden. The blood can be freed from metabolic wastes. Thus, order and normalcy can be restored.

This process is constantly essential because normal blood flow is under continual attack (not only from microthrombi but also cholesterol).

Fibrinolytic enzyme therapy has been used successfully to treat arteries and veins. It can be a lifesaving treatment, with pulmonary embolisms (vessel blockage of the lungs) and myocardial infarcts (blockages of heart vessels). A pulmonary embolism is usually a recently-formed clot.

Certain enzyme mixtures increase the fibrinolytic activity of the blood, and therefore, normalize the hemostatic equilibrium to our body's balancing act.

The deposits of fibrin are dissolved by certain enzymes, the edema is reduced, and the pathologic condition controlled.

Arteriosclerosis

One of the reasons for *arteriosclerosis* is the reduced level of plasmin (a natural body proteolytic enzyme) in the blood to dissolve the continuously deposited fibrin and lipids. The reduced level of plasmin is not sufficient to dissolve the continuously deposited fibrin and lipids. Early intervention with supplemental enzymes (before erosion of the lining) is often effective. Formation of atheroma is avoided.

If deposits within vessel walls are advanced, plaques are already formed, and thus, enzyme mixtures can only partially dissolve them. Despite this, the improvement of blood supply with enzyme therapy causes a continuing reduction of symptoms.

Enzyme mixtures avoid further arteriosclerosis and reduce existing symptoms.

The development of venous disease is based on:

1. the reduction of blood flow velocity;
2. damage to the lining of the blood vessel; and
3. the increased tendency for blood coagulation.

Thrombosis

Systemic enzyme therapy improves the blood fluidity and therefore, improves blood flow.

Pain and swelling are symptoms of inflammation. The causes of pain are chemical and physical irritation of the nerves by toxic, inflammatory products and excessive pressure generated within the swollen tissue. Certain enzyme mixtures eliminate the toxic, inflammatory products. The infiltrated tissue is quickly relieved from fibrinous exudates containing proteins and cells. Swelling is decreased and therefore, the pressure within the tissue is reduced.

Some enzyme combinations can reduce pain and edema. Microthrombosis is caused by an imbalance between the formation and degradation of fibrin. That is to say, the destruction of the sticky substance is too weak. Enzyme combinations have a fibrinolytic effect (they break up fibrin). In addition, they activate the body's own enzymes which also have a fibrinolytic effect. Therefore, fibrinolysis increases to normal levels, resulting in rapid breakup of the thrombi (clots). Hence, the formation of new microthrombi and thrombi is avoided.

Certain enzyme mixtures inhibit thrombi formation and stimulate the destruction of thrombi. Reduction of blood flow decreases the nutrients supplied to the tissues in the area of inflammation. As the microthrombi dissolve, microcirculation improves. With the normalization of the blood supply, there is increased nourishment of the tissue and therefore, healing is stimulated.

In 70 to 90 percent of various venous and arterial vascular diseases, including the post-thrombotic syndrome, healing or marked improvement is noted in such symptoms as: pain, swelling, fatigue, and heavy feeling in the extremities, inflammation, venostasis, microthrombosis, disturbances in blood supply, cold extremities, eczema, and varicose ulcers.

Increased pressure in front of the blockage is a second danger associated with venous thrombosis. The vessels swell and large protein molecules and plasma fluid are forced into the surrounding tissue, resulting in edema (swelling of the legs).

In the beginning, most people see this as just a cosmetic problem. However, appearance is not as important as the health of the leg.

When the protein from the obstructed vein escapes into the surrounding tissue, it changes in character and is treated by the body as an enemy. Around the congested or blocked vessels, the connective tissue cells begin to grow and harden. Further, lymph vessels become blocked, resulting in lymphedema formation.

In the affected region there is inflammation with resulting symptoms of pain, heat, swelling and reddening.

The production of fibrin intensifies.

In this way, the condition intensifies, becomes self-perpetuating, and is known as *post-thrombotic syndrome*. According to conventional medical view, there is not a great deal that can be done.

Post-thrombotic syndrome has been treated successfully throughout the world, using enzyme mixtures. Symptoms of this condition include: leg cramps during the night; a heavy, restless, tense feeling in the legs; and pain in the calves (particularly after sitting or standing for a period of time, and subsiding after lying down or walking).

Chronic Venous Insufficiency

Chronic venous insufficiency symptoms are penetrating and burning pain (primarily at night). The legs feel alternately hot and cold. The ankles swell. You can't stand on your feet very long without becoming fatigued. Poor circulation is an increasing problem (whether sitting, walking, or standing) especially when varicose veins appear.

Leg veins can get some help through: 1) simply elevating your feet, 2) daily gentle massage, and 3) frequently standing on tiptoe.

However, these exercises will not protect veins from becoming damaged. Venous disorders begin by thrombi formation from sluggish blood flow.

The process has gone too far and surgery is required to remove the large clots which settle in the veins. They are too large and cannot be broken down and transported away (no matter how many enzymes are used). The vein has become blocked.

As a result, the blood seeks another avenue. New vessels are formed, bypassing the blocked vessels and allowing venous blood to reach the heart. However, a danger still exists from the thrombus in the blocked vein. The focus of the blocked area still continues to grow in size. The blood that still continues to carry and deposit microthombi presses against the thrombus. Edema can develop in the legs.

Therefore, it is critical to normalize the blood equilibrium and to promote breakdown of the microthrombi, as well. The reparative process is assisted by administration of active enzymes.

Pain in your legs? Everyone will suffer from leg problems sometime during his or her life. It's all too common an occurrence, but one that should be closely observed to rule out serious, life-threatening circulatory disorders.

Harmless looking varicose veins are not particularly dramatic. However, the condition can deteriorate to a deep vein thrombosis or possibly even be the cause of a pulmonary embolism.

Pathologic Venous Processes

In a double-blind cross-over study with ten volunteers, the effect of an enzyme mixture on the blood parameters was studied. Ernst[3] administered thirty tablets of a certain enzyme mixture or placebo per day for a period of two weeks. Then the medication was interrupted for one week, followed by a group change. Results indicated a significant decline in the plasma viscosity after the first and second treatment week. Serum viscosity, enzyme flexibility, and aggregability revealed positive results.

Certain enzyme mixtures induce changes in the blood viscosity in healthy individuals; this quality could be of use in circulatory disturbances.[4]

Certain enzyme mixtures have proven effective in chronic arterial circulatory disturbances, where the swelling condition required treatment.[5] A decrease in swelling was noted and the patients were free of pain while lying down and asleep at night.

Denck[5] gave a certain enzyme mixture to patients. As a result, blood flow was re-established via embolectomy and the ischemia tolerance time increased to over six hours. In addition, post-ischemic swellings and peripheral ischemias occurred much less frequently.

As early as 1962, Doctor J. Valls-Serra,[6] Director, Department of Angiology and Vascular Surgery, University of Barcelona published his experiences with the enzyme treatment of venous inflammation and stated, "Our therapeutic results are excellent, particularly with respect to the duration of therapy. In the case of earlier methods of treatment, it took several months for phlebitides (venous inflammations) to heal. The present methods of treatment bring about healing within only a few weeks. But for post-phlebitic syndrome, too, that is, the continuance of symptoms six months after the appearance of the primary phlebitis, the results were also astonishingly good and superior to all methods of vessel dilation and anti-coagulant therapy previously undertaken."

In 1972, Professor Wolf[6] reported the results of treating 347 American patients having various types of venous disorders and treated with enzyme therapy. Of patients suffering from surface phlebitis, 58 percent were completely symptom-free, 29 percent almost free from symptoms. Only 13 percent exhibited no or only a slight improvement. The treatment duration was considerably less for the enzyme group than the control group. The improvement took place in one half to one third the time needed for the control group.

Doctor H. Denck (Chairman of the Surgical Department of the Municipal Hospital of Vienna) published results similar to those of Wolf.[6] A clinical study using enzyme therapy with post-thrombotic syndrome treatment revealed that 41 percent of the patients had been completely freed from the disorder or were vastly improved after eight weeks, 53 percent were very much better, and only 6 percent reported no change in their condition.

The treatment of venous disorders is not restricted to hospitals. In order to collect the experiences of general practitioners, Dr. Maehder conducted a multicenter study of patients.[6] The results of enzyme treatment on 216 patients having numerous venous disorders were analyzed. Findings indicated 31 percent of the patients had complete recovery, improvement was indicated in 62 percent of cases, and 7 percent remained unchanged.

Enzyme therapy is not confined to those venous disorders, which have already entered the chronic stage. Since it is possible to reduce the incidence of venous disorders as we age, the value of enzyme treatment is even greater.

In his publication on the oral enzyme treatment of inflammatory venous disorders, Dr. Herbert Mahr[6] (Bad Durrheim, Germany) came to the conclusion: "It is important that it is also possible to treat patients at risk (smoking, improper or unbalanced diet, toxins, stress) prophylactically with enzymes. In the case of occupations virtually predestined to suffer venous disorders (salesgirls and all occupations that involve a great deal of standing), such an enzyme preparation can be administered to protect the circulation."

About 60 percent of the German adult population has at least temporary venous trouble.[6] Every eighth adult suffers from chronic venous insufficiency. As a result of venous problems, one to two million people in Germany suffer from nonhealing leg sores. If these statistics are this high in Germany, think how bad it must be in America. These statistics prove the importance of enzyme prophylaxis. In particular, these facts underline the importance of sensible and acceptable preventative treatment for those at risk of thromboses. Further, the connection between thromboses and the risk of cancer is a known fact. Therefore, in treating the one condition, we could be preventing others.

Post-Thrombotic Syndrome

Ninety percent of the time, thrombosis leads to delayed circulatory damage. Post-thrombotic syndrome can result from deep venous thrombosis of the legs. A typical sign of impairment is vein edema.

Undissolved thrombi (clots) can plug up the vessels, impairing the flow. Changes can affect the vein wall and portions of the cup-shaped venous valves. The venous blood return is no longer in one direction, but "pendular" (back and forth).

Collateral leg circulation develops to provide drainage. This causes a dilation of the superficial vessels. The venous valves break down and are unable to properly function, causing varicose veins and resulting in venous obstruction.

The already weakened lymphatic system becomes over-burdened. The capillaries dilate, local tissue and cell pressure increases. Plasma is forced into the tissue spaces. Macrophages and other phagocytes do their jobs, infiltrating the perivascular tissue and chewing up (degrading) the cellular debris.

This is a chronic inflammatory reaction. Characteristic are degenerative disturbances of the skin, pigment, ulcer development, repeated eczema, stasis dermatosis, as well as phlebothromboses and thrombi formation.

However, enzyme therapy is beneficial even in advanced stages, which are usually difficult to treat.

Klein studied enzyme therapy with 100 post-thrombotic syndrome elderly patients (up to ninety-six years of age).[7] After seven to fourteen days, leg swelling was reduced by more than 50 percent. Pain decreased, mobility increased, and no increase in thrombi was noted. Further, there was a decrease in cholesterol and triglycerides, plus an increase in HDL.

With forty post-thrombotic syndrome patients, Morl conducted a double-blind study investigating the therapeutic value of a certain enzyme mixture.[8] The thrombi were three to twelve months old. Patients were tested before the study, after fourteen days, and again at twenty-eight days after the start of therapy. Patients were evaluated subjectively (pain, leg cramps, and fatigue) and objectively (ulcers, skin changes, and edema) in addition to plethysmography measurement of venous pressure, light-reflex rheography and laboratory results.

Lymphedema

Lymphedemas are vascular disorders which can be helped by enzyme therapy. Keim, et al.[9] and Scheef[10] applied proteolytic enzymes systemically, as well as locally. Streichhan and Inderst[11] summarized results of enzyme treatment.

Lymphedema commonly develops following a mastectomy (with excision of axillary lymph nodes) and mammary carcinoma surgery.

Scheef[10] studied enzyme therapy in the prevention of lymphedema following mastectomy. Thirty patients were treated with enzymes for two years. No lymphedema developed in 96.7 percent of the women during the observation period of six years, while only 71 percent of the patients not taking enzymes had no signs of lymphedema. In another study, Konig found similar results.[12]

The Terrible Two: Cancer and Circulation

Research in the area of enzyme therapy and circulatory disorders has revealed some interesting findings. About 100 years ago, the French doctor Trousseau drew attention to the tendency of cancer patients to suffer thromboses and of thrombotic patients to suffer from cancer.[6] Using studies (including dissection statistics from the Hamburg and Munich Pathological Institutes), cancer researchers have confirmed Dr. Trousseau's findings.

The common factor is clear. The cancer cells can multiply undisturbed and form a tumor, by hiding in and under undissolved fibrin.

Effects of Enzymes

Studies indicate enzyme mixtures seem to be effective in treating thrombophlebitis, atherosclerosis, post-thrombotic syndromes, and edema. Enzyme preparations inhibit inflammation in a two-fold manner. That is, in addition to the preparation's own anti-inflammatory properties, the body's natural enzymatic processes are stimulated, without causing immunosuppression. This fact alone presents a considerable advantage over the use of corticosteroids.

Enzyme mixtures increase tissue permeability as well as the rate of degradation of inflammation, toxic products, and necrotic substances. They increase the rate of dissolution of microthrombi, thereby reducing swelling and microcirculation and consequently improving the supply of nutrients and oxygen to the tissue as well as improving the removal of end products of normal metabolism. Therefore, the duration of the inflammatory process is reduced, pain stops more rapidly, and in many instances, wounds heal quickly without scar formation.

ADVANTAGES OF ENZYME TREATMENT:

Rapid effect
Excellent tolerance
No acute or chronic side effects
No risk of hemorrhage
Coagulation controls not necessary
No deferred effects
No suppression of the immune system
Infection defenses are not handicapped
No incompatibility with other medications
Can be used in diabetic patients
Necessity of additional pain medication is reduced
Useful for patients of all ages
Improves the effects of sulfonamides and antibiotics
Useful in post-thrombotic syndrome
Useful in venous and arterial disease
Decreased scar formation with wound healing

Enzyme therapy (hydrolytic enzymes from plants, animals, and fungi) is a proven treatment method for diseases relating to blood vessels.[13] It diminishes the swelling, activates the fibrinolytic system, and stimulates cells such as macrophages, without the long-term side effects of drug therapy.[7,14] The pain and cramps disappear, the swelling goes away, and blood flow increases after a short time. The efficacy of certain proteolytic enzyme mixtures (pancreatin, trypsin, chymotrypsin, papain, bromelain, amylase, lipase, and the bioflavonoid rutin) has been tested by double-blind studies.[7]

Enzymes can't do it alone. Prevention is the word and the Five-Step Jump-Start Plus Enzyme Program is the way. Review the five-step program in chapter 8 and make it part of your longer, happier, more-productive life.

With enzymes and the Five-Step Jump-Start Plus Enzyme Program, you can help fight:

1. Cholesterol in the blood vessels
2. Emboli
3. Thrombi (blood clots)
4. Pathologic venous processes
5. Post-thrombotic syndrome
6. Occlusive arterial disease
7. Thromboembolic complications
8. Lymphedema
9. Cancer from circulatory problems

References

1. *National Vital Statistics Report*, 47(9). Washington, D.C.: Centers for Disease Control, 10 November 1998.
2. *Healthy People 2000*. Washington, D.C.: U.S. Department of Health and Human Services, Public Health Service, 1991, 392–393.
3. A. M. Ernst and A. Matrai, "Orale Therapie mit Proteolytischen Enzymen Modifiziert die Blutrheologie," VI. Kongress der Deutschen Gesellschaft fur Klinische Hamorhealogie, 1989.
4. J. P. Guggenbichler, "Einfluss Hydrolytischer Enzyme auf Thrombusbildung und Thrombolyse," *Die Medizinische Welt* 39:277–280 (1988).
5. H. Denck, "Enzymtherapie in Vascular Disorders," From lecture, Conference on Systemic Enzyme Therapy, Munich, Germany. October 22, 1986.
6. Wilhelm Glenk, and Sven Neu, *Enzyme Die Bausteine des Lebens Wie Sie Wirken, Helfen und Heilen*. Munich, Germany: Wilhelm Heyne Verlag, 1990.
7. K. Klein, "Proteolytisches Enzympraparat Erfolgreich," *Therapiewoche Osterreich* 30:448 (1989).
8. H. Morl, "Behandlung des Postthrombotischen Syndroms Mit Einem Enzymgemisch," *Therapiewoche* 36:2443 (1986).
9. H. Keim, et al., "Methode zur Linderung der Lymphstauung am Arm Nach Behandlung des Mammakarzinoms," *Rontgenberichte* 1:1 (1972).
10. W. Scheef and M. Pischnamazadeh, "Proteolytische Enzyme als einfache und sichere Methode zur Verhutung des Lymphdems nach Ablatio Mammae," *Die Medizinische Welt* 35:1032 (1984).
11. P. Streichhan and R. Inderst, "Konventionelle und Enzymtherapeutische Massnagmen bei der Behandlung Brustkrebsbedingter Armlymphodeme," *Der Prakt. Arztliche* (1991).

12. W. Konig, "Erfahrungen der Robert-Janker-Klinik, Bonn, Mit Systemischer Enzymtherapie und Emulgierten Vitaminen, *Acta Medica Empirica* 37:11 (1988).
13. Michael W. Kleine and Helmut Pabst, "Die Wirkung einer Oralen Enzymtherapie auf Experimentell Erzeugte Hamatome," *Forum Prakt und Allgemeinarztes* 27:42–48 (1988).
14. Rudolf Inderst, "Enzyme Treatment of Vascular Disorders," from lecture, First International Conference on Systemic Enzyme Therapy. September 12, 1990.

CHAPTER 19

Victory over Viruses

FROM MEASLES to herpes, from HIV to the common cold, viruses are everywhere and enzymes may be one of the best ways to control these vicious fellows.

We find viruses almost everywhere on planet earth. But, only a very few are pathogenic (the bad guys). Though viruses are quite small (they can only be seen by an electron microscope), nevertheless, they are responsible for many different diseases in man.

Viruses multiply only after invading a host cell and tricking the cell into reproducing more copies of the virus. Viruses make all the parts they need to reproduce themselves inside our cells. Then, the parts come together and voila a new virus. It's like a car assembly line. At the beginning of the line, all the separate parts are strewn around. Little by little the parts are combined. At the end of the line, all the parts are together in a brand new shiny car. So, assembly is the method by which viruses mature.

In this way, viruses can take over a cell and use it to produce thousands and thousands of tiny viruses (10 to 10,000 times smaller than bacteria).[1] All of the host's energy is spent reproducing viral substances; its normal activities are lost after viral invasion. Our cells are like zombies from the black lagoon doing the master's (virus') bidding. Viruses cause a large variety of diseases in animals and man, such as rabies, influenza, mumps, infectious mononucleosis, polio, measles, chicken pox, hepatitis, HIV, and the common cold, to name a few. Some viruses are known to be cancer viruses and many scientists feel that *all* cancer is caused by viruses. American cancer researchers David Baltimore, Renato Dulbecco, and Howard Temin won the 1975 Nobel Prize for Medicine by demonstrating the interaction of cancer viruses and genetic material within body cells. Cancer viruses find their way into normal cells, making them cancerous. Using viral injections, a number of malignancies can be caused in laboratory animals.

The host cell's protein coat is believed to be important in the attraction, attachment, and actual invasion into the host cell by the virus. Therefore, enzyme therapy seems to have great possibilities in fighting viral disease since enzymes attack the virus protein coat. H. Gotz used proteolytic enzymes against herpes zoster and herpes simplex (fever blister virus) in humans.[2] R. Cleeland completely eliminated the influenza virus with trypsin. Viral diseases have few treatments and are not susceptible to antibiotics. Research in Germany has repeatedly shown that enzyme therapy can provide significant protection against, and treatment of, viral diseases.

WITH WEAK DEFENSES, THE KILLERS ATTACK

Viruses revive and penetrate our healthy cells when our bodies are in a weakened condition. These cells then make copies of the killer invaders. Stress (mental and otherwise), depression, defense-weakening drugs, and illness can all weaken our bodies, leaving them open to viral invasion.

Events which can lead to a reactivation of viral infection might involve the following:

1. Trauma
2. Surgery
3. Menstruation or pregnancy
4. Administration of drugs which weaken the immune response
5. Radiation, such as sun and/or ultraviolet

Herpes

About 90 percent of the populations of Europe and North America are infected with one or more of the six basic types of herpes viruses.

The herpes virus is never actually expelled from our body. Once we have it, it stays in our system, lying dormant, waiting for the next chance to become active.

Whenever our defenses are low (from infection, stress, injury, etc.), the virus can awaken, and wreak havoc on our bodies. Renewed herpes attacks may lead to ulcers and other physical afflictions.

Most of us have had chicken pox, usually when we were children. But did you know that the virus that causes chicken pox is still hibernating in your body and can kick into gear at any time, leading to shingles (herpes zoster)?

When we had chicken pox, our bodies produced antibodies which helped cure the disease. The antibodies, however (along with the virus itself), remain in our system. These antibodies joined with the viral antigens to form "immune complexes."

Herpes zoster (shingles) is this type of a disease. Formed from antibodies and zoster antigens, the immune complexes tend to lodge on the nerve cells. The virus alters the structure of the nerve cells. Immune complexes attached to the nerve activate enzymes triggering a mechanism intended to chew up (lyse) bacteria. This action causes nerve cell damage and the typical symptoms of shingles.

Herpes Zoster (Shingles)

Do you have gastrointestinal disturbances, fever, chills, and malaise with or without pain? These symptoms could relate to many conditions, but specifically, herpes zoster.

Just like any virus, the chicken pox virus can remain dormant in your body, reactivating under times of stress. This can lead to the formation of new immune complexes, which can adhere to nerves, causing inflammation, nerve damage, pain, and the typical symptoms of shingles (blisters, swollen lymph nodes, burning, itching, irritation, and general fatigue). Nerve pain (especially running from between the ribs to the spine) can be extremely serious. The most frequent areas of dry skin and scab eruption are found between the ribs and usually only on one side. If the lesions and incredible pain last more than two weeks, there is quite possibly an underlying immunologic defect or malignancy; see a doctor. Luckily, most patients recover without residuals, except for occasional scarring of the skin.

Until recently, scientific texts on the treatment of shingles agreed that treating the cause of the disease was impossible. Though vitamin B_{12} or gamma globulin was frequently prescribed, these medications were of little help. Just as with a cold, the only thing to do was wait, since shingles should heal itself in two or three weeks. This treatment plan was great unless post-zoster neuralgia developed. Post-zoster neuralgia can cause excruciating nerve pains lasting for years, or even a lifetime, and is resistant to current medical treatment.

Herpes may occur at any age, but is most common after fifty. Immunosuppressive drugs are usually the therapy of choice, but after treatment, the body's immune system is weakened and open to viral and bacterial invaders. It's like taking from Peter to pay Paul. Everyone is different, but the repeated use of immunosuppressive drugs can weaken our body. In addition, we can become weakened from many different outside influences, such as disease, drugs, mental stress, depression, and daily stresses. In this weakened condition, viruses revive and penetrate the healthy cells, producing replicas of the invading killer cells. Ultimately, cancer and other diseases can develop.

The use of immunosuppressive drugs can be compared to using a two-edged sword. In fighting off certain conditions with these drugs we might, at the same time, be impairing the work of our immune system and weakening ourselves to future invaders.

Since this is an inflammatory condition (affecting the sensory root ganglia and skin), proteolytic enzymes have been quite effective. In addition, enzyme therapy is effective because enzymes act against immune complexes.

Genital Herpes (Herpes Simplex II)

The estimated annual incidence of symptomatic genital herpes is 200,000 cases, with total estimates of infection as high as thirty million cases in the United States alone![3] These high levels are expected to continue, with recurrences among more people and increasingly serious complications for pregnant women and newborns. Efforts to control genital herpes are currently hampered because as many as three-fourths of the people having the disease *don't know they have it* and, therefore, don't take precautions to keep from spreading it to others. Of those who are aware they have the condition, only about 8 percent have mild symptoms and 2 percent are seriously ill.[4]

Symptoms can include:

1. Gingivostomatitis of the skin and mucosa, which lasts for seven to twenty-four days (until it has completely healed)
2. Infection of the cornea and conjunctiva of the eye
3. The mucosa of the vagina or the glans penis may be affected

After a short time, clusters of small pustules appear and fill with clear fluid. The skin is slightly raised and has a moderately inflamed surface, frequently accompanied by severe itching. To make matters worse, there is no known cure for genital herpes.

How Enzymes Can Fight Viruses

Professor Max Wolf discovered a connection between viruses and enzyme therapy while treating cattle for papillomatosis, which he felt was caused by a virus. He then used enzymes to treat bean plants which he subsequently infected with the tobacco mosaic virus. Those plants treated with the enzymes resisted the virus, while those not treated became infected. Enzymes have also been effectively used in the treatment of equine viral cough and bovine influenza, two conditions deadly to animals.

Enzymes work on viral infections primarily because they eliminate the immune complexes, thus they can help eliminate inflammation, pain and fatigue. Several researchers around the world have been successful using enzymes to fight herpes.

In 1965, a Dr. Dorrer of Germany administered an enzyme mixture to twenty-four patients with shingles. Beginning within twenty-four hours of the first appearance of the blister, he gave the patients massive doses of enzyme. He found that their pain ceased within three days and blisters formed scabs more quickly than normal. One important finding of his study was that no patients later suffered post-zoster neuralgia.[5]

In 1968, Dr. Bartsch began using large doses of enzymes to treat a group of twenty-three cancer patients, suffering from herpes zoster.[6] A second group was treated with the standard medication prescribed for the disease. The superiority of the enzyme therapy became so obvious to Dr. Bartsch that he abandoned the standard treatment before its completion. He could not justify the use of customary methods when it was obvious the enzymes were vastly superior.

According to Dr. Bartsch, "At this time we regard the treatment of herpes zoster with proteolytic enzymes as being the therapy that is most effective, most free from side effects, and the best."[6]

Dr. Wolfgang Scheef has studied enzymes in the treatment of herpes zoster and believes they should be applied immediately after the blisters appear. So far, all patients receiving the treatment have improved, and none developed post-zoster neuralgia. Most impressive is that all affected patients became free of pain within fourteen days.[7]

We know that warts are caused by a virus. In Germany, enzyme treatment of newly formed warts usually involves local application of enzyme salve, plus tablets, taken orally.

Pneumonia, influenza, colds, and other viral conditions appear to be successfully treated by enzyme therapy. The course of these conditions can be considerably shortened. Symptoms appear milder and complications are thus avoided.

Another consideration in fighting viruses is increased proteolytic enzyme concentration in our blood plasma. If we take proteolytic enzymes on a maintenance basis, we are fighting respiratory infections in two ways: first, we are fighting its viral effect, but secondly, its inflammatory effect. Regarding inflammation, it is generally accepted and experimentally demonstrated that a healthy mucous membrane gives us greater protection against most bacterial invasions. However, an inflamed or irritated area opens the doors for pathogenic germ invasion. We close the doors with anti-inflammatory enzymes.

You can see the importance of staying healthy and following a preventive maintenance care program. This is why the Five-Step Jump-Start Plus Enzyme Program is so vital. When we are run down and dragged out, it is more difficult to protect against germs. Enzymes deprive the bacteria of their breeding and hiding places.

VICTORY OVER VIRUSES

A quick review on viruses and enzyme therapy reveals some very important points:

1. Viruses are almost everywhere.
2. All viruses are alike in that they have protein coats containing nucleic acid. This nucleic acid is composed of DNA and RNA (the building blocks of life), making it possible for the virus to reproduce itself.
3. Enzymes fight viruses by breaking up this protein coat.
4. The antibodies in our system bind to viral antigens creating immune complexes.
5. These immune complexes can float in the bloodstream and become lodged in the tissues, leading to inflammation, pain, and sometimes, chronic diseases.
6. Enzymes have the ability to break up these immune complexes, enabling them to be swept from the body by the defense system.
7. Enzymes are able to stimulate the immune system so that it can do the job for which it was designed, that is, protecting us from viruses and disease.

Therefore, these viral vixens can be controlled through enzymes. Viruses might be tiny, but our enzymes can search them out and control them. Who knows? Enzymes might even prove to be the answer to the common cold.

References

1. E. Michaud and A. Feinstein (Ed.) *Fighting Disease, the Complete Guide to Natural Immune Power*. Emmaus, PA.: Rodale Press, 1989, 308.
2. Francis X. Hasselberger, *Uses of Enzymes and Immobilized Enzymes*. Chicago: Nelson Hall Publishers, 1978, 123.
3. Centers for Disease Control, Office of Women's Health, *Sexually Transmitted Diseases*. Retrieved from http://www.cdc.gov/od/owh/whstd.htm.
4. W. Konig, "Enzyme therapy for virus infections and carcinoma," *Acta Medica Emperica* 1–9 (1991).

5. Wilhelm Glenk and Sven Neu, *Enzyme Die Bausteine des Lebens Wie Sie Wirken, Helfen und Heilen*. Munich, Germany: Wilhelm Heyne Verlag, 1990.
6. W. Bartsch, "Proteolytic Enzymes in the Treatment of Herpes Zoster," *Der Informierte Arzt.* Jahrgang 2/nr 10 Oktober. 1–7. 1974.
7. Wolfgang Scheef, "Enzymtherapie, Lehrbuch der Naturheilverfahren," *Hippokrates* Bd. II, S:95–103 (1987).

CHAPTER 20

New Hope for AIDS

WHAT IF you're diagnosed with HIV or AIDS? After the initial shock, depression, and fear have passed, you may ask, "What can I do about my problem?" Enzymes are the answer, no contest!

The problem is extremely serious. According to the Centers for Disease Control, an estimated 1.5 million Americans were infected with the Human Immunodeficiency Virus (HIV).[1] New studies show there is hope for AIDS patients. However, the key to unlocking the door to longer life is *not* orthodox treatment, but alternative health care.

AIDS refers to a group of serious conditions (including lymphoma, wasting syndrome, and Kaposi's sarcoma) which apparently are the result of HIV infection. The AIDS definition was expanded by the Centers for Disease Control and Prevention (CDC) on January 1, 1993 to include laboratory evidence of severe immune suppression (including pulmonary tuberculosis, recurrent bacterial pneumonia, and invasive cervical cancer), *when found in people also having HIV infection*. Without the presence of HIV, the condition is not diagnosed as AIDS, according to the CDC.

The confusion in the definition of AIDS is loaded with controversy. In the summer of 1992, at least forty adults were reported to have "full-blown AIDS" (especially pneumocystis carinii pneumonia [PCP] and Kaposi's sarcoma with characteristic AIDS immune markers) without *ever* having had either HIV antigens or antibodies. Is it AIDS or isn't it? How can someone having all the symptoms of AIDS (yet not having HIV) *not* be classified as having AIDS (by the CDC)? This is a puzzlement.

Why do some AIDS patients live longer than others? While there are individual differences in long-term survival, there are also common threads weaving through each life; courage, the guts to face almost anything, and the ability to laugh at one's self. Seven individuals followed by *Parade Magazine* have lived an average of more than eight years after AIDS diagnosis, the majority having stepped away from drugs, taken charge of their lives, adopted a natural food diet (including a supplement program), joined support groups, and having a will to live productively.[2]

Holistic and integratively minded physicians and researchers are making the greatest strides against AIDS, says Dr. Michael Culbert of Bradford Research Institute (BRI).[3] Further, BRI argues that HIV alone may not be the cause of AIDS and that various cofactors (such as infections, environmental causes, etc.) must be present. The individual may be predisposed to these cofactors because of damage to his or her defense system.

This damage can result from mental stress, drugs (both street and prescriptions), a chemically altered food supply, fluoridated water, nutrient imbalances, immunizations/vaccinations, low-level electromagnetic emissions, prior or ongoing viral/bacterial/venereal infections, yeast infections, or parasites.[3]

Effective AIDS treatment can never be achieved by seeking a "magic bullet." Rather, it necessitates an integrated program of body detoxification, diet control, balanced body chemistry (enzymes, vitamins, minerals, amino acids, hormones, plus other nutrients), herbs, plus nontoxic therapies against random viruses, bacteria, yeast, parasites, and mycoplasmas.

According to Dr. Culbert, long-term AIDS survivors have three things in common:

1. Either complete abandonment of "orthodox" allopathic medicine, utilization of an "alternative" approach, or a combination of the two;
2. They eat right; and
3. They think right.

Sounds like the Jump-Start Program to me.

The majority of people with HIV become deficient in several important nutrients early in the course of their disease. Dr. Joan

Priestley studied the immunomodulating effects and clinical benefits of nutritional therapy on 164 HIV-positive patients (average twenty-three months since HIV diagnosis). For an average of eighteen months, all subjects took therapeutic doses of nutritional supplements.

From the study, Dr. Priestley concluded that nutrient replacement therapy appeared to significantly influence maintenance of stable body weight in HIV/AIDS patients. Even when CD4 lymphocyte counts dropped below 50, well over half (58 percent) of Priestley's patients survived for at least another year. Finally, survival rates appeared to be higher among those who had never taken AZT as opposed to those who had taken AZT.[4]

Dr. Ian Brighthope of Australia successfully manages patients with AIDS and suggests the following "Micronutrient Supplementation Program": diet, rest, meditative techniques and positive imagery; vitamins, minerals, evening primrose oil, echinacea; mistletoe; thymus extract; Lactobacillus acidophilus tablets or powder; Nystatin; Ketoconazole; garlic or garlic extract; Pao D'arco tincture; beet and carrot juice; metallo proteins; licorice root extract; homeopathics; aged aloe vera juice; and pancreatic enzymes. (For further information, see *The AIDS Fighters*, Ian Brighthope, M.D., with Peter Fitzgerald, from Keats Publishing, Los Angeles, 1988.

New enzyme studies in the United States, Costa Rica, the Czech Republic, Germany, and other countries give promising results.

In 1991, Dr. G. Ramirez (University of Costa Rica) studied HIV patients for twelve months and found a combination of enzymes (including pancreatin, papain, bromelain, lipase, amylase, trypsin, chymotrypsin, plus rutin), taken orally, increased CD4 cells, while symptoms (temperature, night sweats, serious diarrhea, nausea, anorexia, cough, skin rash, difficulty breathing on exertion, palpable lymph nodes, lymphadenopathy, encephalopathy, nephropathy, and retinopathy) improved by 83.6 percent.[5] Opportunistic infections (such as Pneumocystis carinii pneumonia, toxoplasmosis, candidiasis, viral infections, gastrointestinal infections, and other infections) improved 84.0 percent, while autoimmune diseases reduced measurably.

Open clinical pilot trials were used to study HIV infected patients (stages one to five according to the Walter Reed classification).[6–8] Patients experienced an improvement in the general condition with unchanged CD4 and CD8 counts, as well as a reduced total number of typical HIV clinical symptoms. So far, most of Dr. Jager's experiences were achieved using three doses per day of ten enterically coated tablets of a certain proteolytic enzyme mixture, or a measuring spoon (granulate) of that enzyme mixture, three times daily.

Both Bastyr and National Colleges of Naturopathic Medicine have developed successful programs for HIV/AIDS patients. Bastyr College studied thirty men with AIDS-related complex (ARC-CDC Class IV-A) for one year. Results indicated that naturopathic treatment may help slow the progression of AIDS and provide relief for patients with mildly symptomatic immune deficiency.[9]

As is the case in every other viral infection, a person with AIDS produces specific antibodies. These antibodies are directed against the HIV/AIDS virus and can be detected in the blood. Because they can be detected in the blood, they are important for diagnosis. If these antibodies are present, the person is "HIV positive." This does not mean, however, the he or she is already suffering from AIDS.

Many years can pass before the inactive HIV virus (present in the body) becomes activated. A stress on the body, such as the administration of immunosuppressive drugs (which depress the body's own defenses) can set in motion the disease process, leading to full-blown AIDS (and usually death). If you are on Azathioprine, or Cyclophosphamide (given to transplant as well as rheumatoid arthritis patients), you are on an immunosuppressive drug.

Though immunosuppressive drugs are capable of inhibiting the spread of the AIDS virus, there is a very high price to pay for reducing viral proliferation. These drugs depress your defenses, leaving your body helpless to fight every other infection. For this reason, such drugs should only be used in cases where the immune system is already completely weakened and further weakening is impossible.

It should be clear, too, that viral infections should be combated with treatment programs reinforcing the body's immune system. In order to successfully oppose the disease, the body must marshall all the methods available.

This is accomplished by relieving the body of metabolic poisons, by a sensible life style and diet, and by the use of biological therapies aimed at strengthening the body. This therapeutic approach must be modified on an individual basis.

The aim of therapy should be to stimulate the virus-inhibited activity of the immune system's helper cells, especially the macrophages. The inhibition of macrophages and prevention of virus destruction (by macrophages) is a result of immune complex formation (from the AIDS virus).

What makes AIDS so dangerous is the fact that the immune complexes (with the virus antigens), are not degraded. The AIDS virus attacks and paralyzes the defense cells. In addition, the HIV antigens (locked in the immune complexes) react to and bind with the helper cells. The complement, alarmed by this process, destroys the helper cells (which are coupled to the HIV viruses). The complement now sees the helper cells as the enemy.

More and more scientists confirm that the destruction of helper cells caused by immune complexes is a primary reason for the deadly consequences of AIDS.

Enzyme therapy is an effective treatment because it mobilizes the immune complexes in the tissues (bringing them into the circulatory system and available for enzymatic degradation). This is probably the most effective natural health care approach in the treatment of AIDS. The immune deficiency is reduced and further progression of AIDS symptoms is halted. The patient remains HIV positive, but can become symptom-free. As yet, it is as impossible to remove all HIV viruses from the body as it is to eliminate any other viruses.

Numerous hospitals and medical centers in America, Germany, and France are presently applying this principle in AIDS patient treatment.[10]

According to Drs. Glenk and Neu, ". . . there is nothing in this life that cannot be improved, reinforced, or healed through the administration of enzymes." This might seem like a strong statement. However, if all activities in the body require some form of enzymatic activity, this statement has value. It seems that enzymes can give us not only a healthy life, but also a longer life.

ENZYME THERAPY IS EFFECTIVE AGAINST AIDS

1. Enzyme treatment significantly limits the progression of HIV, in the early stages.
2. Patients' symptoms are greatly improved.

3. Enzyme therapy can normalize suppressed low-grade helper-cells.
4. Enzyme treatment can delay onset of disease symptoms with HIV-positive individuals. This delay appears to continue indefinitely, in some cases.
5. Infectious diseases and possibly, malignant diseases, can be decreased in number. This is possibly due to increased macrophage activity (the PacMen of our bodies), improved helper cell defense, and accelerated degeneration of foreign cells.

KEY POINTS IN FIGHTING AIDS

Since alternative health care is demonstrated to be more effective than orthodox drugs, why aren't the media and AIDS organizations talking more about these successes? By their silence, by *not* exposing the general public to alternative health care, are they allowing AIDS patients to die without a fighting chance?

It seems obvious that hope is on the horizon. Certain AIDS patients are living longer, and the majority are following an alternative health-care program. The drive to live, an unwavering positive mental attitude, and a positive self-image, along with a commitment to a daily lifelong therapy program seem critical for success. It is essential that body, mind, and spirit be in balance—for the health of it!

References

1. Centers for Disease Control. Atlanta, Georgia.
2. Bernard Gavzer, "What Keeps Me Alive," *Parade Magazine*, January 31, 1993, 4–7.
3. Michael Culbert, personal communications, 1993.
4. Joan Priestley, Posterboard presentation at the VIII Intl. Conference on AIDS/III Std. World Congress, 1992.
5. Gerhard Stauder, "Intermediate Evaluation of an Open Clinical Trial," October 15, 1991.
6. H. Jager, "Hydrolytic Enzymes in the Treatment of Patients with HIV-Infections," from lecture, First International Conference on Systemic Enzyme Therapy. September 11, 1990.

7. Gerhard Stauder, D. Fuchs, H. Jager, W. Samtleben, and H. Wachter, "Adjuvant Therapy of HIV Infections with Hydrolytic Enzymes: Course of Heopterin, CD4-T-Cells, Immune Complexes (IC), and Clinical Efficacy," 8th International Workshop on Biochemical and Clinical Aspects of Peridines, St. Christoph (Tyrol), February 11–18, 1989.
8. Gerhard Stauder, Karl Ransberger, Peter Streichhan, W. Van Schaik, and W. Pollinger, "The Use of Hydrolytic Enzymes as Adjuvant Therapy in AIDS/ARC/LAS Patients," *Biomedicine and Pharmacotherapy* 42:31–34 (1988).
9. Leanne Standish, et al., "One Year Open Trial of Naturopathic Treatment of HIV Infection Class IV-A in Men," *The Journal of Naturopathic Medicine* 42 (1992).
10. Wilhelm Glenk and Sven Neu, *Enzyme Die Bausteine des Lebens Wie Sie Wirken, Helfen und Heilen*. Munich, Germany: Wilhelm Heyne Verlag, 1990, 118.

CHAPTER 21

The Scars of Multiple Sclerosis

BACK IN the early 1950s, I dated a young lady with multiple sclerosis (MS). At that time, the diagnosis of MS seemed like a death knell.

Although I knew this young lady (let's call her Mary) for only a year, it was understood that she would never become serious with anyone as it was only a matter of time before she would be gone from this world.

Although bright, attractive, and very healthy looking, her condition noticeably deteriorated in the year I knew her. She dragged her left foot more and more (occasionally, she tripped as she walked). Her right leg was often numb. Sometimes her walk was unstable and she staggered as though she had had too much to drink; but she didn't drink. The muscle-wasting became increasingly more evident. Her speech sometimes slurred, her vision often blurred and she saw double. Although she was always happy on the outside, there was a constant atmosphere of impending doom.

In those days, there were no well-known champions for the MS patient (like Dr. Roy Swank, Dr. John McDougall, or Dr. Christa Neuhofer), no knights in shining armor mounted on white horses to fight the relentless dragons of neuromuscular deterioration, pain, and despair. No doctors seemed willing to risk their careers to seek an alternative to the dead-end drug approach in fighting this dreaded disease.

Mary's only therapy was a prescription of daily drugs. Her only promise—despair and death. It was Mary's never-ending faith that kept her going.

Eventually, I transferred to another school. Mary and I lost touch. But, I have always wondered what happened to her.

What if you are diagnosed with MS? Luckily, therapy choices have improved and you need not receive a death sentence similar to Mary's. After the initial shock and fear have passed, you ask, "What is MS and what can I do about it?"

The words multiple sclerosis mean "many scars." It is a slowly progressing disease, characterized by disseminated patches of demyelination in the central nervous system. The nerves of the central nervous system (the brain and spinal cord) are covered with an insulation called *myelin*. For unknown reasons, in MS the myelin breaks down, interrupting signals which normally travel between adjacent nerve fibers. This results in multiple and varied neurologic symptoms and signs usually with exacerbations and remissions (symptoms come and go, similar to rheumatoid arthritis).

This disease is very difficult to accurately diagnose because it mimics many other neuromuscular disorders. Symptoms usually relate to the site of demyelination. Frequent symptoms are clumsiness or weakness of a hand or leg; paresthesia (burning, prickling) in one or more extremities; and visual disturbances, such as diplopia (seeing double), dimness of vision, and so forth.

The course of MS may vary considerably. It is either characterized by episodes (with more or less complete remission of the symptoms between episodes), or it can be chronic progressive (the most severe form which leads to death within a few years). Alternatively, the disease may start in episodes and become chronic progressive over a period of time.

What Causes Multiple Sclerosis?

As yet, there is no agreement on what causes lost myelin function, increased inflammatory response, and nerve destruction, randomly affecting the function of various organs.

Why does this happen? Nobody knows for sure, though there are many theories. One theory is the "virus" theory, that viruses are possibly involved in myelin disturbances. Viruses may lie dormant in the body for years before reappearing and damaging the myelin. Other theories include: the autoimmune theory (for an unknown reason, the body attacks its own myelin) and the chemical theory (some MS patients have chemical encephalitogenic antibodies in their systems—but it is unknown which came first—the chemical or the disease). The genetic predisposition theory assumes that some genetic error increases an individual's risk of developing MS. It is possible that MS is *genetically inherited*, but it is difficult to explain what might be genetically altered. It has also been suggested that improper diet, or a selenium deficiency (for instance) might be the trigger. Selenium is a nonmetallic element not created in the body, but absolutely necessary for many metabolic processes. We acquire it through the food we eat, the air we breathe, and absorb it through the pores in our skin. We rid ourselves of it through perspiration, breathing, urine, and stools. Selenium is an incredible element, both an enzyme activator and an enzyme inhibitor.

Dr. Swank's theory is that capillaries feeding the myelin cells become plugged because of sticky blood platelets—possibly a reaction to malutilization of fats leading to disturbed uptake of unsaturated fatty acid. This theory stems from the ratio of saturated to unsaturated fatty acids in everyday diet and its correlation to the demyelination of the central nervous system. Other theories include Dr. David Horrobin's theory of prostaglandin imbalance; and the theory that an emotional shock precedes the onset of many cases of MS. As you can see, there are many theories on the origin of MS. It is possible that some or all of them are correct to a varying degree.

One interesting note: MS occurs more often in temperate climates (between 40 and 60 degrees north and south latitudes—similar to wide belts circling the globe). The closer an area to the equator, the fewer the cases of MS.

Whatever the cause, MS is neither preventable nor curable, at this time. However, treatment can help alleviate some of the symptoms. Two-thirds of MS patients remain functional after twenty years, and 75 percent may never need a wheelchair.[1]

Multiple sclerosis is a devastatingly insidious monster which affects between 250,000 to 350,000 inhabitants of our great country.[2] Each year, over 8,000 Americans are struck down.[1]

Women are twice as susceptible as men, with Caucasians being the predominantly involved race. MS becomes most apparent between thirty and fifty years of age, but seldom strikes those under fifteen or over fifty.[1]

How Do People Treat MS?

Everyone has his or her own theories as to how best to treat MS. For instance, in his book, *New Hope, Real Help For Those Who Have Multiple Sclerosis*, John Pageler describes how he found himself struck down with this devastating disease. (The book is available from Mr. Pageler at 6200 102nd Terrace North, Pinellas Park, FL 34666.) This courageous man's fight for survival is an example for us all. John feels that strict adherence to his diet is a major factor in his successful MS program. He says you must avoid hard fats and sweets with equal determination and develop new eating habits, including supplementation.

John believes supplements are necessary to rebuild the damaged area of myelin or to possibly construct new pathways around the damaged areas.

He feels you must be strong enough to look your favorite aunt in the eye and say, "I can't eat the meal you have planned for a week, and spent two days preparing, because it doesn't conform to my diet."

John believes the MS patient can't expect immediate improvement in his or her health condition. "We didn't get sick with MS in a few months, and we won't improve our situation in a few months either," he says. "But, as a patient living day to day with your disabilities, you will know when the first improvement happens." In John's case, it took him six years to go from typical MS functional blindness, numbness of the left-side extremities, slurring of speech, cane and leg braces, plus lack of bladder control, to feeling normal enough to play tennis and be able to return to a job as a radio announcer.

Currently, high dosages of cortisone therapy (1,000 milligrams prednisolone per day with successive dosage reduction to zero) are a favored procedure by some. Azathioprine is another drug used to treat this disease. Both these approaches supposedly curb excessive autoimmune activity. The disadvantages are obvious. First, they suppress the immune system. Secondly (like using a sledgehammer to crack nuts), they render patients susceptible to infections which severely aggravate the existing symptoms. Furthermore, according to Dr. Christa Neuhofer of Salzburg, Austria who has treated MS patients for over twenty years, cortisone therapy ceases to be effective after three to four treatments.

Azathioprine has also proved disappointing with long-term treatment. Although symptoms may improve over the first two years (in 30 to 40 percent of the patients), after five years Azathioprine's effect will diminish and in the long run, no change in the progression of MS has been recorded.[3] In addition there is the threat of side effects such as impaired hematopoiesis (blood cell formation and development), cholestasis (suppression of bile flow) and hepatitis (inflammation of the liver). Prolonged therapy increases the risk of neoplasias (abnormal tumor growth).[3]

What a dreary outlook for MS patients. But there may be hope on the horizon.

According to independent studies in Greece, the Czech Republic, and the United States, the number of circulating immune complexes (CIC) in serum blood levels of MS victims is consistently greater than those of normal, healthy individuals.

These findings conclude that MS may depend upon, or be caused by, immune complexes and is probably an autoimmune disease. Could it be possible that factors previously theorized, such as genetic predisposition, selenium deficiency, fatty acid imbalance, and viruses all lead to an immune system malfunction and an uncontrollable reaction to the body itself? Sounds good to me.

Over twenty years ago, Professor Max Wolf told of frequently occurring remissions following enzyme therapy in patients with multiple sclerosis.[4] In 1973, Professor Tsiminakis, Institute of Neurology of the University of Athens, and Dr. Sedivi, Hospital of Neurology, (the former) USSR, reported similarly good results with certain enzyme mixtures in the treatment of MS.[4]

At that time, the working mechanism, or a connection between immune complexes and MS was entirely unknown. Professor Wolf was of the opinion that the antiviral effect of enzymes could be responsible for their success in treating MS.[4]

We cannot yet say what causes multiple sclerosis. However, we do know that immune complexes (of a particular size, type, and concentration) make MS worse. The removal of these pathogenic immune complexes provides relief (interrupting the progressive myelin deterioration and even reversing the function loss of individual nerves). Further, research has shown that circulating immune complexes can be reduced through the use of orally-ingested proteolytic enzymes.[5]

Dr. Neuhofer[6] has extensive experience in the treatment of MS. Also, she is an MS sufferer who (personally) has limited the progress of this dreaded disease through the use of certain enzyme mixtures.

In 1972, Dr. Neuhofer[4] began treating MS patients with enzyme mixtures, initially using them on herself (as her first MS patient). Her improvement was dramatic (particularly the symptoms affecting the eyes, bladder, and rectum).

Two hundred of Dr. Neuhofer's MS patients were followed in a clinical research study.[7] Findings indicated that symptoms of the eyes which measurably improved under enzyme therapy, were partial blindness, eye pain due to retrobulbarneuritis of the optic nerve, double vision, turbidity, and paresis of the eye muscles. Other symptom improvements were increased sphincter control of the bladder and of the rectum. Subsequently, a large number of MS patients were treated by Dr. Neuhofer with these enzyme preparations.

Dr. Neuhofer and her colleagues have treated over 350 MS patients with enzyme therapy.[8] In 1986, she published the statistical results of enzyme treatment on 150 patients who had been treated for over two years.

Her patients were instructed to follow a diet based on whole food (with a very high level of raw foodstuffs containing polyunsaturated fatty acids).

A number of the parameters investigated included: enzyme dose level, type of preparation, what form, and time of administration.

Multiple sclerosis is a disease that is very individual in its expression. There is virtually no uniform disease pattern. Patients differ in their symptom complexes as to manifestation and course of action. Therefore, the whole art of treating MS with enzymes is dependent on a number of considerations relating to the patient's individuality (on taking the right dose levels at exactly the right time). The treatment program is based on whether the disease progresses in a stepwise fashion (with intervals of at least four weeks between acute episodes), whether an initially stepwise progression converts to a chronic status (uniformly developing disorder), which nerves have already stopped functioning, and how rapidly the disease is progressing (how long it has been present).

A question of great importance is how long the condition has existed and the type of prior treatment (what drugs were used, doses and length of time). It is important to know if drugs which suppress the immune system (such as cortisone-containing medications) have been taken. These drugs influence the degree to which the patient's immune system has been weakened.

In her study, a total of 107 patients suffered from chronic progressive MS. Of these, enzyme treatment lead to an appreciable improvement in forty-five patients and the progression was at least diminished in twenty-six patients. In addition, twenty-four patients discontinued their treatment because their health insurance would not bear the treatment costs. Deterioration of condition occurred in only twelve patients. All patients whose conditions continued to deteriorate had previously been treated with a drug (Azathioprine) which destroys the body's own immune system.[8]

But, what is the effect of the orally administered and/or injected enzyme mixtures on progressively deteriorating MS patients? It is known that deterioration is accompanied by an increase in the level of immune complexes. Can the enzyme therapy exert a delaying effect?

Dr. Neuhofer's results were significant.[8] Statistics reveal that of forty-three patients suffering this stepwise progressive deterioration, thirty-five exhibited appreciable improvement and, in some cases, the abolition of all symptoms of paralysis. The conditions of the remaining eight patients were at least stabilized (i.e., their condition did not continue to deteriorate).

Not one single patient was observed to suffer a deterioration of symptoms during therapy. There were no serious side effects.

Since these initial studies, other doctors have adopted Dr. Neuhofer's therapy regime. Many have reported similar positive results. Closer investigation of the disappointing cases revealed that the treatment program required for a certain case had not been accurately followed. For instance, when MS progresses, it is essential that enzyme injections be initiated at the *first signs* of a new exacerbation. If this is neglected for even one or two days, it is impossible to affect the deterioration phase.[5,8]

In any case, it is evident that there is a relationship between the concentration of immune complexes and the flare-up phase of multiple sclerosis. As we know, there are periods of waxing and waning. During these bouts, more than 80 percent of the patients exhibited blood serum with CIC-complexes and highly elevated IC-titers.[9]

Hope for MS Sufferers Reviewed

At present, the therapeutic goals in treating MS should be to: 1) alleviate existing symptoms; 2) stop the episode if possible; 3) reduce the number of episodes; and, 4) check the progress of the disease.

In addition, the patient should follow a modified Five-Step Jump-Start Plus Enzyme Program with special emphasis on controlling his or her environment, both external and internal.

THE FIVE-STEP JUMP-START PLUS ENZYME PROGRAM:

1. Detoxify (detoxification, fasting, juicing). Eliminate all food additives, preservatives, refined carbohydrates, as well as "hard fats."
2. Eat a well-balanced diet. A good diet for MS patients would be either the Swank Diet (Swank, R. L., and Dugan, B. *The Multiple Sclerosis Diet Book*. Doubleday, N.Y., 1987) or the McDougall Plan (McDougall, J. A., and McDougall, M. A. *The McDougall Plan*. New Win Publishing, Inc. Clinton, N.J., 1983). It is important to change eating habits and eliminate (or reduce) animal proteins, refined carbohydrates, stimulants, and all refined, canned, frozen, or otherwise processed food or drink. Eat fresh fruits, vegetables, and grains.
3. Use enzyme, vitamin, and mineral supplements.
4. Exercise daily.
5. Have a positive mental attitude and reduce stress. Stress (both physical and emotional) will exacerbate MS symptoms.

References

1. Personal communication, Multiple Sclerosis Foundation. Letter dated July 28, 1992.
2. National Multiple Sclerosis Society Information Resource Center and Library, Compendium of Multiple Sclerosis Information, 1997.
3. Dr. Ulf Baumhackl (St. Polten) presentation at 17th Symposium in Vienna, 7 Dec., 1991 Systemic Enzyme Therapy, Current Position and Recent Advances.

4. Karl Ransberger and W. van Schaik, "Enzymtherapie bei Multipler Sklerose," *Der Kassenarz* 41:41–45 (October 1986).
5. Wilhelm Glenk and Sven Neu, *Enzyme Die Bausteine des Lebens Wie sie Wirken, Helfen und Heilen*. Munich, Germany: Wilhelm Heyne Verlag, 1990.
6. C. Neuhofer, in: Wrba, H. "Systemic Enzyme Therapy, Newest Status and Progress," *Therapie Wochenschrift* 7:37 (1987).
7. Karl Ransberger, personal communication. 1991.
8. C. Neuhofer, "Enzymtherapie bei Multipler Sklerose," *Hufeland Journal Biologisch-medizinisches Zentralorgan* (1986).
9. C. Neuhofer, "Multiple Sclerosis: Treatment with Enzyme Preparations," from lecture, First International Conference on Systemic Enzyme Therapy (September 12, 1990).

CHAPTER 22

Enzyme Grab Bag

THERE ARE a multitude of uses for enzymes, a few of which we have touched on in the earlier chapters of this book. But enzyme treatment can be used both systemically and locally. It can act in pancreatic insufficiency as an enzyme substitute, in local thrombolysis, chemonucleolysis (with chymopapain), in debridement of ulcers, and as an ointment for burns.[1] Enzyme therapy can go so far as to work in treatment of systemic fibrinolysis, acute and chronic inflammation, circulatory, traumatic and rheumatic conditions, in viral diseases, and in disease prevention.[1] Some other applications include scars, wounds, cataract surgery, and gingiva therapy in dentistry.[2]

SINUSITIS

Sanders and Taub[3] conducted a double-blind study of sixty people to determine the effectiveness of ananase on chronic sinusitis sufferers. Ananase (bromelain) is a proteolytic enzyme known for its anti-inflammatory and anti-edema activity. There was marked relief in a large majority of the patients after the sixth day of treatment. Nasal breathing had become less difficult and the redness and edema of the nasal passages receded. Headaches became less pronounced and nasal secretions more liquefied.

Another study involved bromelain therapy in forty-eight patients with acute sinusitis or nasal mucosal inflammation, nasal secretions, breathing difficulties and associated headache.[4] Mucosal inflammation was reduced in 83 percent of those on bromelain, as compared to 52 percent of those receiving the placebo. Breathing difficulty was relieved in 78 percent of the patients on bromelain, as compared to 68 percent of those on the placebo. Headaches, too, improved in the bromelain group (87 percent) compared to the placebo group (68 percent).

NEUROLOGICAL ASPECTS

Not only can enzymes be used in the treatment of physical impairments and disabilities, but research has shown some remarkable correlations between enzyme levels and neurological disorders. Specifically, a study was conducted to investigate the renin-angiotensin system in melancholia and its relationship to hypothalamic-pituitary-adrenal (HPA) axis function.[5] There was significantly lower angiotensin-converting enzyme activity (ACE) in melancholia patients than normal. Because angiotensin stimulates the sympathoadrenal system and HPA-axis activity, reduced ACE activity could be involved in a down-regulation of both hyperactive systems in melancholia.

Other neurologic disorders under observation with respect to enzymes, are depression, schizophrenia, and obsessive-compulsive behavior (such as anorexia nervosa and bulimia).

CATARACT SURGERY

Proteolytic enzymes could reduce the occurrence of damage during cataract surgery. In cataract formation, the crystalline eye lens or its capsule becomes opaque. Cataract surgery involves removal of this opaque eye lens. In order to do this, the lens must be detached from the ligaments (zonules) which hold it. Tearing or ripping can result. This tearing can delay healing and even damage other segments of the eye. Complications are reduced if the ligaments are exposed to a dilute solution of chymotrypsin for two to four minutes after the incision has been made in the cornea.[2]

The lens is removed more easily after the protein-dissolving enzyme selectively breaks down the ligaments. Evidently, other parts of the eye are not damaged by the chymotrypsin. The patient usually regains vision (after the lens is removed) by using eyeglasses.

It has even been suggested that enzymatic treatment might be the solution to the staggering problem of cataracts in Third World countries.[6] It is estimated that eighteen million people living in Third World countries are blind because of cataracts. This is simply too many people to cure through surgery. However, enzymes could dislocate the cataractous lens and—with the help of inexpensive eyeglasses—could solve this enormous problem.

Tooth Plaque and Decay

Tooth decay is a gradual pathologic dissolution and disintegration of tooth enamel and dentin with eventual deterioration of the pulp. Tooth decay is the most common human disorder, except for the common cold.

Dental plaque is a combination of salivary glucoproteins, desquamated mucosal cells, and polysaccharides. Plaque serves as a localized site of acid production.

Refined carbohydrates, frequently consumed, are implicated in the formation of dental caries (cavities). Between-meal snacks, especially of sucrose-containing foods, make it worse. Nonsticky foods (such as a fresh apple) are potentially less harmful than sticky foods (such as a dried apple).

If plaque is not flushed away by the action of the saliva (or oral musculature), dental caries may begin on the exposed root and tooth surfaces. Saliva's remineralization effect and buffering capacity are its main roles in preventing caries.

If not removed, plaque's soft sticky deposit accumulates on the surface of the teeth. Even on clean teeth, it builds up in about twenty-four hours. Most of us visit the dental hygienist who cleans our teeth, removing plaque buildup. We also use special toothpaste and mouth rinses to fight "the grungies." But did you know enzymes are effectively used in oral hygiene?

Many plaque-controlling preparations include combinations of proteases, cellulases, amylases, and lipases in various combinations. The consensus of opinion seems to be that those preparations containing a high percentage of proteolytic enzymes are the most effective against plaque formation. Several toothpastes that contain enzymes are now on the market. Even an infant toothpaste is now available. This product, called First Teeth™ (manufactured by Laclede Professional Products, Inc.), contains lactoperoxidase and lactoferrin and is designed to reduce the bacteria that cause gum irritations and early plaque buildup. Enzymes, when added to mouthrinses and dentifrices, can also reduce your risk of gingivitis.[7]

Proteolytic enzymes can even help reduce the distress patients experience after wisdom tooth extraction. If a wisdom tooth has to be surgically removed because of partial or complete impaction, the patient tends to suffer a range of problems after surgery. These problems could include difficulty in swallowing, locking of the jaw (almost total inability to open the mouth), misalignment of the lower jaw towards the operated side when opening the mouth, and a surgically-caused painful swelling of the mucosa in the area of the surgery.

Low-grade temperatures and increased acute phase protein levels are the body's postoperative response. Further, there is the increased risk of an infection.

The experience of Dr. Kurt Vinzenz of Vienna, Austria that proteolytic enzymes greatly alleviate the postoperative problems in wisdom tooth extractions was confirmed in a randomized, double-blind study on eighty patients.[8] Proteolytic enzymes were administered from the first postoperative day. As a result, the jaw lock and swallowing problems disappeared more quickly. By the fifth postoperative day, the midline deviation had almost returned to normal, while in the placebo group, the jaw alignment still deviated by 2.5 millimeters at the end of the study (seven days after surgery).

According to Dr. Vinzenz, these results could be improved further by starting the enzyme medication several days prior to surgery. Due to the difficulty in swallowing on the day of surgery, these symptoms could be greatly reduced by increasing this dosage two days before the proposed tooth extraction.

Further, the effectiveness of the enzyme, papain, in treating swelling and other complications of dental surgery was studied by Dr. G. D. Magnes.[2] He performed a double-blind study and found papain significantly reduced swelling, pain, and other signs of inflammation. His conclusions were that papain was an effective and safe adjunct therapy in reduction and control of swelling, pain, and inflammation as a result of oral surgery trauma.

SIDESTEPPING PENICILLIN'S SIDE EFFECTS

Side effects and severe reactions to penicillin are common. However, an enzyme, *penicillinase*, which changes penicillin to penicilloic acid (which is nonantigenic and nonantibiotic), was released in 1958.

Fever, hives, pain, and hardness at the site of the injection, and sometimes, anaphylactic shock are typical side effects of penicillin.

Symptoms disappear as penicillinase destroys the causative factor, penicillin. In over 80 percent of cases treated, penicillinase has favorably changed the course of penicillin, from shock-like syndrome and life-threatening anaphylactic shock to simple hives (urticarias).

Penicillinase removes the irritant (penicillin, which is an antigen). However, it does not repair the harm done in the antigen-antibody reaction. We can use the example of a burn to clarify pencillinase's action. When the enzyme removes the agent causing the burn (penicillin), the effects of the burn still remain.

THE ARTIFICIAL KIDNEY AND ENZYMES

Since blood-clotting is a major problem in the artificial kidney, enzymes streptokinase and urokinase have been used to dissolve the clots. Further, because waste product removal (such as creatinine, urea, and phenols) is a problem in kidney dialysis fluid, Dr. Thomas M.S. Chang of McGill University in Montreal, Canada used an "artificial cell" of microencapsulated urease and an absorbent for ammonia to remove these waste products.[2]

PANCREATIC INSUFFICIENCY

In addition to insulin, the pancreas is responsible for producing many of the enzymes you need for digestion. However, enzyme production can be inhibited if the pancreas is damaged or diseased, or if the duct running from the pancreas to the small intestine is blocked for any reason. Pancreatic enzyme products have a long history of safe use in the treatment of exocrine pancreatic insufficiency and have been available in the United States since about 1938.

Some digestive products should be taken with meals so as to buffer them against stomach acid. However, because each product is formulated differently, follow label directions. Some patients may experience better results with certain products than with others.

PROCTOLOGY

After surgery for hemorrhoids, the use of analgesics is usually unavoidable since the anal region is very pain-sensitive. Bowel movement regulation must be endured during the first two weeks, postoperatively.

It appears that proteolytic enzymes (in ointment form) might have an application in this area.

Actually, enzymes work in at least three ways: 1) they decrease the time of inflammation; 2) they stimulate and assist the immune system; and 3) they also work as digestive aids, breaking down the food particles and facilitating digestion.

As with any condition, it's a job for super enzymes and the Five-Step Jump-Start Plus Enzyme Program.

IN REVIEW

There are countless applications not discussed in this book. The list is endless. You, the reader, are referred to this book's bibliographies and to the many libraries sprinkled throughout this vast world of ours.

Greater life and energy are within your grasp. Only you can "prevent forest fires" and only you can wake up to make your dreams come true for better health. *Take charge of your life! If you don't, no one else will!*

References

1. Karl Ransberger, "Introduction to Treatment with Enzyme Combinations: the Development of this Therapy, its History and Current Status," from lecture, First International Conference on Systemic Enzyme Therapy. September 11, 1990.

2. Frances Hasselberger, *Uses of Enzymes and Immobilized Enzymes*. Chicago: Nelson-Hall, 1978.
3. S. J. Taub, "The Use of Ananase in Sinusitis—A Study of 60 Patients," *The Eye, Ear, Nose, and Throat Monthly* 45:96–97 (1966).
4. R. E. Ryan, "A Double-Blind Clinical Evaluation of Bromelains in the Treatment of Acute Sinusitis," *Headache* 7:13–17 (1967).
5. Michael Maes, S. Scharpe, et al., "Lower Angiotensin I Converting Enzyme Activity in Melancholic Subjects: A Pilot Study," *Biological Psychiatry* 32(7):621–624 (1992).
6. L. J. Girard, "Dislocation of Cataractous Lens by Enzymatic Zonulolysis: A Suggested Solution to the Problem of the 18 Million Individuals Blind from Cataracts in Third-World Countries," *Ophthalmic Surgery* 26(4):343–345 (July–August 1995).
7. M. Midda and M.W. Cooksey, "Clinical Uses of an Enzyme-Containing Dentifrice," *Journal of Clinical Periodontology* 13(10):950–956 (November 1986).
8. Kurt Vinzenz, presentation at the 17th Symposium in Vienna, 7 Dec., 1991, Systemic Enzyme Therapy: Current position and recent advances.

CHAPTER 23

Our Future with Enzyme Therapy

WELL, FRIENDS, we have walked the tightrope of life together. Through this book we have seen how to execute the fine balance of walking on the tightrope of life—without falling off before our time. The aim of this book has been to give you a jump-start each day—a jump-start for life by improving your knowledge of enzymes and enzyme therapy.

TAKE CHARGE OF YOUR LIFE

Take charge of your body, your mind, your whole being. Don't let others control your life or health. With this in mind, a true physician should be a teacher, a guide, helping you to make decisions which will improve your style of life and your health.

The Five-Step Jump-Start Plus Enzyme Program is a guide you can implement to make positive, yet simple, lifestyle changes.

The very knowledge that you can control what is happening in your life can improve your energy and your enzymatic activity.

The way to keep healthy and energized is to follow the Five-Step Jump-Start Plus Enzyme Program:

1. Flush out the garbage from your body (detoxification).
2. Eat an enzyme-rich, well-balanced "Jump-Start" diet, emphasizing fresh foods.

3. Use enzymes and other supplements to jump-start your life.
4. Exercise daily (walk, swim, run, aerobics, etc.).
5. Have a positive mental attitude. You are a very special person.

Since Dr. Beard first used enzyme therapy to treat cancer patients at the turn of the century, we've come a long way in enzyme therapy. Let enzymes work for you. Enzymes have a wide variety of advantages without serious long-term side effects—and enzymes are far less expensive than those costly drugs.

Let's briefly review some of the major applications of enzyme therapy:

1. Energizer
2. Helps with all metabolism
3. Digestive aid
4. Detoxifier
5. Vitamin and mineral helper
6. Slows aging
7. Better skin, fewer wrinkles
8. Weight loss
9. Faster response from injury and inflammation
10. Chronic condition buster
11. Immune system booster
12. Fights arthritis
13. Helps bad backs
14. Helps fight cancer
15. Helps fight many women's diseases
16. Circulation rejuvenator
17. Victory over those vicious viral vixens
18. New hope for AIDS
19. Fights multiple sclerosis

That's impressive! Still, enzyme therapy is in its infancy. No longer are enzymes used just as aids to digestion. Enzyme therapy's future is closely associated with the continued deciphering of the immune system. We already know a great deal, but every new discovery gives more insight into this essential, complicated, and multifaceted system. An intact immune system depends on an optimal supply of enzymes. By no means has every question been answered, every experiment completed, or every problem solved, but researchers are opening new doors every day.

Suffering millions will someday receive relief and be freed from previously devastating conditions such as rheumatoid arthritis, AIDS, most chronic kidney inflammations (of immune origin), chronic progressive liver inflammation, allergies, pancreatitis, intestinal disorders (Crohn's disease and ulcerative colitis), clogged arteries, nerve tissue disorders (such as multiple sclerosis), lung disorders, and many other debilitating conditions.

Don't get sick in America! That's the cry heard throughout our great land. If the disease doesn't kill you, the hospital and doctors' costs will. America is a sick society. We're run down, dragged out, and headed to the glue factory.

Prevention is the answer! Health and prevention of disease is the most cost-effective health insurance we can buy.

It's not only what we make that counts, it's what we don't spend. If we're healthy, we don't get sick, don't have to pay a doctor, buy drugs, or pay hospital charges. Plus, we feel great, have more energy, and look *marvelous*.

Folks, we're in a survival mode. Not only are we being killed on the streets of America, we're being killed from within by our polluted bodies. We are dying in our own fat, our own waste, and our own pollution.

Let's put order back into our lives and take chaos out. Let's take back our streets and our bodies. Let's take charge of our lives. Smokey the Bear doesn't have to be the only one "preventing." Our own forest fires are our runaway chronic killer diseases. You know what Ben Franklin said, "An ounce of prevention is worth a pound of cure." Take these words to heart.

This book has given you a road map to follow. Now, it's up to you. Use the book, refer to it. Wear it out. Buy more and give to friends and relatives. Show them you care. This book should start you on your voyage. It's not intended to be an end, but merely a beginning, a stimulant, a friend, a resource.

POSTSCRIPT

Each day brings new discoveries, new applications. As your author traveled through the enzyme adventure, he was awestruck by the varied actions, applications (both actual and potential), and power of these wondrous enzymes. It was difficult to end this book because practically each day brings new discoveries. Therefore, this book does not represent the final word about enzymes, but merely a definitive beginning.

Appendix A

Naming Enzymes

The names of most enzymes end in "ase." The first enzyme name (proposed in 1833) was *diastase*.[1] It was later suggested that all enzymes be named by adding the suffix, "-ase" to the root indicative of the nature of that particular enzyme's substrate. Though no longer named this way, most enzyme names do end in "-ase" with the exception of a few (e.g., papain, trypsin, pepsin, and chymotrypsin).

Classification of Enzymes

Enzymes are classified according to their substrates and the nature of the reaction they catalyze. There are six main groups of enzymes with each group having fundamentally different activities. The six groups are oxidoreductases, hydrolases, transferases, lyases, isomerases, and ligases.

Oxidoreductases play an important role in metabolism. They are concerned with oxidation-reduction processes. Oxygenation may be thought of as adding one oxygen or removing two hydrogen atoms from a molecule, while reduction is just the opposite process. It is impossible for a molecule to be oxygenated without the simultaneous reduction of another molecule.

Hydrolases catalyze reactions whereby chemical bonds are broken with the addition of one water molecule (that is, hydrogen [H] is added to a fragment molecule and hydroxide [OH] is added to the other). Included in this group are proteolytic, lipolytic, and amylolytic enzymes.

Transferases catalyze reactions in which substances other than hydrogen are transferred (best known are the transamidases, which are of real value in medical diagnostics).

Lyases are those that rearrange molecular structures, converting a molecule into a related molecule which is said to be an isomer of the original molecule. A conversion of a D-form to an L-form would be an isomerization.

Isomerases catalyze the interconversion of aldose to ketose sugars.

Ligases catalyze the formation of a chemical bond between two molecules.

Enzymes are large molecules. However, in order to get an idea of relative size, let's compare a molecule of the enzyme trypsin with the size of man. If trypsin (one of the first enzymes discovered) were magnified in size to 3.9 inches (about ten centimeters) and compared with a man enlarged in the same proportion, he would be 24,800 miles tall (about 40,000 kilometers); tall enough to wrap himself around the entire world at the equator.

LOCK AND KEY

A biochemist would say that each enzyme is substrate-specific, though there are exceptions to every rule. It is not very far from the truth, though, to say that each enzyme type is only able to take up and alter one specific species of substrate in its precisely shaped active site.

Each enzyme is also effect-specific. This means that it can only carry out one quite specific change on the substrate (i.e., produce one single effect).

One of the early biochemists, Professor Fischer, illustrated this beautifully with the example of the lock and the key.

The lock will only disengage when the specific shape of the key fits exactly into the keyhole of the lock. The substrate is the key. In addition, the unlocking process can only work in one particular way (the key must be turned either to the left or to the right).

Another theory for enzymatic action is the "induced-fit theory." This theory postulates that the substrate must do more than simply fit into the already preformed shape of the active site. The theory states that the binding of the substrate to the enzyme must cause a change in the shape of the enzyme that results in the proper alignment of the catalytic groups on its surface. This concept has been likened to the fit of a hand in a glove, the hand (substrate) induces a change in the shape of the glove (enzyme). Though some enzymes function according to the lock and key theory, most apparently function according to the induced-fit theory.

Substrates

Every living organism contains a variety of different biochemical structures, which are known as *substrates*. The substrates are components necessary for the processes of life. They dash here and there. In so doing, they come in contact with an enzyme. The substrate is drawn into the enzyme's active site. If they fit exactly in this precisely-shaped site, a reaction occurs. For one tiny moment, the substrate and enzyme form a unit. This enzyme has been produced for a specific biochemical reaction, which now takes place. When the substrate is large in size, it is surrounded by a series of enzymes and these enzymes alter the substrate bit by bit. Substrates are biochemical factories carrying out work on an enzymatic conveyor belt.[2]

Most enzyme activities involve cleavage (splitting) of a substrate. Only about three to five percent of enzymes put something together (synthesize instead of cleaving). These are the "anabolic" enzymes and not the cleaving "catabolic" enzymes. When cleavage takes place, the substrate which fits into the active site is broken up. The broken-up pieces are then released in two portions. One product is used to produce new substrates (after the product is broken down into its biochemical components). The other product is another substrate which now searches for an enzyme (undergoing yet another change). This process continues until a product is created which has a particular function within the organism.

The length of time required to attract a substrate to enzyme active sites, then use and release the substrates once again, is dependent on a number of variables.

Each enzyme has developed its own working speed and particular working conditions. We can get some idea of the speed at which an enzyme works by considering the laziest known enzyme (lysozyme). Lysozyme helps destroy bacteria. Lysozyme manages to process about thirty substrate molecules per minute. That is one substrate every two seconds. This contrasts with the fastest worker of all, carboanhydrase, which processes a fantastic thirty-six million substrate molecules in one minute.[2]

The speed of substrate transformation, however, is not the same as the intensity of activity. Basically, it has to do with the work conditions.

In addition, an enzyme makes its working mood dependent on whether there is a large amount of substrate waiting to be changed and whether a great deal of product has built up. (The more substrate, the higher the activity of the enzymes. But the more product, the lower the enzyme activity.)

The activity of an enzyme is not active in the usual sense. An enzyme remains an unaltered protein body. However, it is the presence of the enzyme which brings about a particular effect.

Every protein changes in time and does not live forever. It ages and dies. Since enzymes are proteins, they also age. Eventually, the active site ceases to be a true template and errors occur. When an enzyme begins to exhibit signs of wear and tear, another enzyme comes along and makes short work of it. The worn-out enzyme is degraded, dissolved, and transported away.

Some enzymes only have a life of about twenty minutes. After this, the enzyme is replaced by a newly produced enzyme of the same type. Other enzymes remain active for several weeks before they are replaced as part of the aging process in life's cycle.

One of the most fascinating properties of all enzymes is their ability to work with each other, to form into cooperatives (when this is necessary), and to continually exchange information with other enzyme cooperatives. Thus, the enzymes maintain harmony within the processes of life. The equilibrium of all systems and the common effort for a common goal are all positive properties.

This process cannot be brought about by any single enzyme. Each enzyme cannot just go ahead with its work and hope that everything will be fine. Once we understand how all enzymes are able to work together for the common good, we might be able to develop the ideal form of bodily functions. Enzymes often work consecutively (in steps) in order to carry out important tasks in the body and to maintain the systems in an ideal balance (i.e., between surplus and shortage). These linear systems are known as *enzyme cascades.*

In enzyme cascades, one enzyme activates the next enzyme, which then activates another enzyme, until finally, the last enzyme triggers the desired effect.

One of the reasons for this process is economy, as such small steps require far less energy than one larger, more complex step. Safety is another reason. Blood coagulation and liquefaction are examples. The contraction and dilation of the blood vessels or the triggering and activation of defense forces are other examples.

It is essential for the body to maintain a narrow optimum in all these activities. It simply must not go to extremes. The results of faulty circulatory functions are arteriosclerosis or hemophilia. If our defenses do not function properly, we could be destroyed by the next attack of pathogens.

Enzyme Inhibitors

This is the reason for the versatility of enzymes, but it can also be very dangerous if the enzymes are out of control. For this reason, our bodies are not only equipped with the necessary enzymes, but also with a double safety system to prevent these enzymes from activating at the wrong time.

Since protein is one of the main body components, why don't the protein-degrading organisms simply destroy us? We know that pepsin digests food proteins in the stomach. Why doesn't it digest the stomach (also made of protein)?

The answer is the body's double safety system. First, the enzymes, which continually produce replacements, are not initially active. Enzymes possess (at some point along the amino acid chain) special amino acids which block their activity (just like the safety catch on a pistol).

These harmless enzymes float throughout the blood and lymph systems. Only when a particular enzyme reaction is required does another (specially-activated) enzyme remove the safety catch. Only then is the enzyme ready and able to change any substrate it meets. This is the first safety system.

Secondly, enzyme inhibitors constitute a safety system. These enzyme inhibitors (produced by the body itself) are able to occupy the enzymes' active sites and thus put these sites out of action when too many active enzymes are present.

Some enzyme inhibitors remain attached for the rest of the enzyme's life, others detach themselves (thereby becoming active once again).

Numerous other substances exist which can be used to neutralize quite specific enzymes and thus deliberately interfere with metabolic processes. One of the most famous medications in the world, aspirin, works according to this principle. Aspirin was administered in enormous quantities for decades without any known rationale for its effects.

Aspirin consists of acetylsalicylic acid. This acid is a foreign substance that can attach itself to an enzyme (cyclooxygenase). This enzyme plays a role in blood coagulation and inflammation. In this way, the aspirin inhibits blood coagulation and the blood becomes more fluid. Also, it inhibits inflammation (resulting in pain reduction).

Other examples of this process would be antibiotics such as penicillin or steroids. The desired mechanism of action is always inhibition of a particular enzyme activity.

References

1. *The New Encyclopedia Britannica*. Chicago: Encyclopedia Britannica, Inc. William Benton, Publisher, 1974, 896–902.
2. Wilhelm Glenk and Sven Neu, *Enzyme Die Bausteine des Lebens Wie Sie Wirken, Helfen und Heilen*. Munich, Germany: Wilhelm Heyne Verlag, 1990.

Appendix B

GLOSSARY

AIDS Acquired Immune Deficiency Syndrome.

Amine An organic compound containing nitrogen.

Antibodies A modified type of serum globulin synthesized by lymphoid tissue in response to antigenic stimulus.

Anticoagulant Any substance which inhibits, suppresses, or delays blood clotting.

Antigen A high molecular weight substance or complex, usually protein or protein-polysaccharide complex in nature, which, when foreign to the bloodstream of an animal, on gaining access to the tissues of such an animal, stimulates the formation of a specific antibody and reacts specifically *in vivo* or *in vitro* with its homologous antibody.

Arteriosclerosis Hardening and thickening of the arteries, with loss of elasticity.

Arthrosis A disease of a joint or articulation which limits its motion.

B cells One of the three branches of the body's white cells. They are bone marrow-derived and produce antibodies. They are also called lymphocytes.

Bekhterev's disease Ankylosing spondylitis; chronic inflammation of the spinal column.

Cancer A cellular tumor the natural course of which is fatal and usually associated with formation of secondary tumors.

Complement One of the serum enzyme systems. Its functions include mediating inflammation, rendering bacteria and other cells susceptible to phagocytosis, and causing membrane damage to pathogens.

Cortisone A carbohydrate-regulating hormone from the adrenal cortex, which is concerned with glyconeogenesis.

Crohn's disease Chronic inflammation of the small intestine.

Double-blind study A study in which neither the participants nor the doctors administering the medication are aware whether the placebo or a preparation is being given.

Embolism A sudden blood clot in a blood vessel.

Enzyme From Greek, meaning "in yeast." An organic compound which accelerates or produces a catalytic action.

Enzyme cascade The complement components interact with each other so that the products of one reaction form the enzyme for the next, thus a small initial stimulus can trigger an increasing cascade of activity.

Exudate Fluid which escapes from blood vessels (usually during inflammation).

Fibrin A whitish, insoluble protein formed from fibrinogen by the action of thrombin (fibrin ferment) as in the clotting of blood. Fibrin forms the essential portion of the blood clot.

Fibrinolysis The splitting up of fibrin by enzyme action.

Herpes Zoster (Shingles) A virus related to chickenpox. Symptoms: severe pain; sometimes chills and fever, malaise, meningismus. Signs: crops of clear vesicles (blisters) along course of a cutaneous nerve; tender regional lymph nodes.

Immune complex Antigen bound to antibody is termed an immune complex, but in the context of immunopathology complexes may also include complement components.

Immunosuppression The artificial suppression of the body's immune system.

Inflammation The response of tissue to injury, with the function of bringing serum molecules and cells of the immune system to the site of damage. The reaction consists of three components: 1) increased blood supply to the region, 2) increased capillary permeability in the affected area, and 3) emigration of cells out of the blood vessels and into the tissues.

Lupus erythematosus An autoimmune disease which leads to changes in the skin, joints and internal organs.

Lymphatic system The name for all the organs of the immune system. It is made up of two subdivisions: primary (or central) and secondary (or peripheral) organs.

Lymphocyte A variety of white blood corpuscle.

Macrophages One of the three branches of the body's white cells. Metchnikoff's name for a large mononuclear wandering phagocytic cell which originates in the tissues. Fixed macrophage: a quiescent phagocyte, such as the histiocyte of loose connective tissue or those lining the sinuses of the liver, spleen, lymph glands, and bone marrow. Free macrophage, an ameboid phagocyte present at the site of inflammation; called also polyblast and inflammatory macrophage.

Multiple sclerosis Spotty demyelinization of nervous system, cause unknown. Symptoms: visual disturbances, incoordination, myasthenia, paresthesias, loss of sphincter control. Signs: personality changes, ataxia, dysarthria, intention tremor, ocular palsies, retrobulbar neuritis, hyperactive deep reflexes, diminished abdominal reflexes, trophic changes in skin.

Natural Killer (NK) **Cells** Are able to kill foreign cells through direct contact, by producing a cytotoxin, or cell poison.

Pancreatitis Inflammation of the pancreas.

pH (Potential Hydrogen) A 15-step scale measuring acidity or alkalinity.

Phagocyte Any cell that ingests microorganisms or other cells and foreign particles. In many cases, but not always, the ingested material is digested within the phagocyte.

Phagocytosis The engulfing of micro organisms, other cells, and foreign particles by phagocytes.

Phlebitis Inflammation of a vein.

Phospholipids Lipids containing phosphorus, which on hydrolysis yields fatty acids, glycerin, and a nitrogenous compound. Lecithin, cephalin, and sphingomyelin are the best known examples.

Prostaglandin (PG) A group of cyclic fatty acids that possess diverse and potent biologic activities affecting cell function in every organ system. Prostaglandin synthesis is inhibited by aspirin, indomethacin, and other nonsteroidal anti-inflammatory agents. Metabolism takes place mostly in the lungs, renal cortex, and liver. Metabolites are excreted in the urine. At high concentrations prostaglandins are considered anti-inflammatory because they ameliorate experimental adjuvant arthritis in animals. PGs may play an important role during systemic, as well as local, inflammatory reactions.

Proteases A general term for a proteolytic enzyme.

Proteolysis The hydrolysis of proteins into proteoses, peptones, and other products by means of enzymes.

Proteolytic enzyme An enzyme that promotes proteolysis.

Rheumatoid arthritis A disease of unknown cause, it is characterized by proliferative inflammation of the body's connective tissue, especially the articular synovia.

Serotonin A constituent of blood platelets, enterochromaffin cells, and of other organs; used as an experimental agent to induce vasoconstriction and alter neuronal function.

Stasis A stoppage of the flow of blood or other body fluid in any part.

T cells One of the three branches of the body's white cells. They are thymus-derived, aka: lymphocytes.

Thrombosis Obstruction (partial or complete) of a blood vessel by a blood clot.

Ulcer A sore forming on a tissue wall because of loss of protective covering on that wall.

Ulcerative colitis Chronic inflammation of the lining of the large intestine with ulcerative wall destruction.

Vasculitis Inflammation of a vessel.

Vasodilation The dilation of the local blood vessels caused by the action of mediators on the smooth muscle of the vessel walls, producing increased blood flow.

Appendix C

In order to keep you abreast of current happenings in health, the following is a group of excellent health-related magazines and newspapers:

Better Nutrition
6151 Powers Ferry Road, N.W.
Atlanta, GA 30339

Energy Times
P. O. Box 649
Melville, NY 11747-9806

Health Counselor
Impakt Communications, Inc.
P. O. Box 12496
Green Bay, WI 54307–2496

Health News & Review
A quarterly health news paper dedicated to a better life for you and your family through preventive health care and regular exercise programs.

Let's Live Magazine
320 N. Larchmont Blvd., Third Floor
Los Angeles, CA 90004

Longevity Magazine
1965 Broadway
New York, NY 10023–5965

New Age Journal
42 Pleasant St.
Watertown, MA 02172

Nutrition Science News
1301 Spruce Street
Boulder, CO 80302

Prevention Magazine
P. O. Box 7319
Red Oak, IA 51591–0319

Total Health Magazine
1650 North 100 East, Suite 2
St. George, UT 84770-2505

Townsend Letter for Doctors and Patients
911 Tyler St.
Port Townsend, WA 98368–6541

Yoga Journal
P.O. Box 469088
Escondido, CA 92046

Resource List

A number of companies manufacture and distribute enzyme products. The following is a partial list of some of these companies. The author and publisher do not endorse these companies but list them here for the reader's convenience. Great care has been taken to ensure that the addresses and phone numbers are correct; however, we cannot be responsible for typographical errors or changes in information.

ADH Health Products, Inc.
215 N. Route 303
Congers, NY 10920
(914) 268-0027
Fax: (914) 268-2988

ADM Laboratories, Inc.
American Desert Manufacturing
5536 W. Roosevelt Street, Suite #1
Phoenix, AZ 85043
(602) 272-3777
Fax: (602) 272-1500

Advanced Labs
8759 Airport Road, Suite C
Redding, CA 96002
800-955-5553
Fax: (916) 223-0699

Advanced Nutrient Science
P. O. Box 668
1157 North Hotsprings Drive
Parker, CO 80134
(303) 840-0555
Fax: (303) 840-0550

AIM (American Image Marketing)
3904 East Flamingo Avenue
Nampa, ID 83687
(208) 465-5116
Fax: (208) 463-2187

AkPharma, Inc.
P. O. Box 111
Pleasantville, NJ 08232
(609) 645-5100
Fax: (609) 645-0767

Alltech, Inc.
Biotechnology Center
3031 Catnip Hill Pike
Nicholasville, KY 40356
(606) 885-9613
Fax: (606) 885-6736

Amano Enzyme U.S.A. Company, Ltd.
1157 North Main Street
Lombard, IL 60148
800-446-7652
Fax: (630) 953-1895

American Biologics
1180 Walnut Avenue
Chula Vista, CA 91911
(619) 429-8200
Fax: (619) 429-8004

American Dietary Labs
14631 Best Avenue
Norwalk, CA 90650
800-423-8837

American Health
4320 Veterans Memorial Highway
Holbrook, NY 11741
800-445-7135
Fax: (516) 244-1777

American Laboratories, Inc.
4410 S. 102nd Street
Omaha, NE 68127
(402) 339-2494
Fax: (402) 339-0801

Anabolic Laboratories, Inc.
17801 Gillette Avenue
P. O. Box C19508
Irvine, CA 92713
(714) 863-0340
Fax: (714) 261-2928

Archon Vitamin Corporation
209 40th Street
Irvington, NJ 07111
(201) 371-1700
Fax: (201) 371-1277

Arve's Zyming Beauty Products
P. O. Box 1869
Flagler Beach, FL 32136
(904) 439-3305
Fax: (904) 439-7303

Aveda
4000 Pheasant Ridge Drive
Minneapolis, MN 55449
800-283-3224
Fax: (612) 783-4110

Barth's Nutra Products Corp.
3890 Park Central Blvd., North
Pompano Beach, FL 33064
800-645-2208
Fax: (954) 978-7093

Bayer Corporation
P. O. Box 3100
Elkhart, IN 46515
800-248-2637

Bio-Cat, Inc.
Route 2, Box 1475
Troy, VA 22944
(804) 589-4777
Fax: (804) 589-3301

Bioenergy Nutrients, Inc.
6395 Gunpark Drive, Suite A
Boulder, CO 80301
800-553-0227

Bio-Nutritional Products
41 Bergenline Avenue
Westwood, NJ 07676
800-431-2582
Fax: (201) 666-2929

BioSan Laboratories, Inc.
P. O. Box 325
8 Bowers Road
Derry, NH 03038
(603) 432-5022
Fax: (603) 434-4736

Biotec Foods
1 Capitol District
250 S. Hotel Street, Suite 200
Honolulu, HI 96813
800-331-5888
Fax: (808) 529-9342

Biotics Research Corporation
P. O. Box 36888
Houston, TX 77236
(713) 240-8010

Body Mechanics
624 Estuary Drive
Bradenton, FL 34209
800-264-1114

Botanicals International
2550 El Presidio Street
Long Beach, CA 90810-1193
(310) 637-9566
Fax: (310) 637-3644
E-mail: botan@botanicals.com

Canadian Natural Products Ltd.
B15.60020 2nd Street, SE
Calgary, Alberta
Canada T2H 2L8
(403) 252-0177
Fax: (403) 252-0176

Carlson Laboratories, Inc.
15 College Drive
Arlington Heights, IL 60004
800-323-4141 or (708) 255-1600

CC International, Inc.
P. O. Box 2452
Rancho Santa Fe, CA 92067
800-775-3575
Fax: (619) 756-1334

C. C. Pollen Company
3627 East Indian School Road
Suite 209
Phoenix, AZ 85018
800-875-0096 or (602) 957-0096

Cell Tech
1300 Main Street
Klamath Falls, OR 97601-5914
(503) 882-5406
Fax: (503) 844-1869

Country Life
101 Corporate Drive
Hauppauge, NY 11788
(516) 231-1031
Fax: (516) 231-2331

Crystal Star Herbal Nutrition
4609 Wedgeway Court
Earth City, MO 63045
800-736-6015

Dr. Goodpet Labs
332½ AE Beach Avenue
P. O. Box 4728
Inglewood, CA 90309
800-222-9932
Fax: (310) 672-4287

Dr. Grandel, Inc.
626 W. Sunset Road
San Antonio, TX 78216
(210) 829-1763

Doctor's Pride, Inc.
75 Bi-County Boulevard
Farmingdale, NY 11735
800-645-9909
Fax: 800-397-4252

Douglas Laboratories, Inc.
600 Boyce Road
Pittsburgh, PA 15205
888-DOUGLAB
Fax: (412) 494-0155

Dr. Enzyme's Health Products
P. O. Box 92094
Portland, OR 97292-2094
(503) 256-9901

Earthrise Company
424 Payran Street
Petaluma, CA 94952
(707) 778-9078
Fax: (707) 778-9028
E-mail:Info@Earthrise.com

Enzymatic Therapy, Inc.
825 Challenger Drive
Green Bay, WI 54311
800-558-7372
Fax: (920) 469-4400
Internet: http://www.enz.com

Enzyme Development Corporation
2 Penn Plaza, Suite 2439
New York, NY 10121-0034
(212) 736-1580
Fax: (212) 279-0056

Enzyme Process International
2035 East Cedar Street
Tempe, AZ 85281
800-655-9092 or (602) 731-9290

Ethical Nutrients
971 Calle Negocio
San Clemente, CA 92673
800-668-8743
Fax: (714) 366-2859

Fruit of the Land
14631 Best Avenue
Norwalk, CA 90650
800-423-8837

Futurebiotics, Inc.
145 Ricefield Lane
Hauppauge, NY 11788
(516) 273-6300
Fax: (516) 273-1165

Garden State Nutritionals
100 Lehigh Drive
Fairfield, NJ 07004
800-526-9095
Fax: (201) 575-6782

General Nutrition, Inc.
921 Penn Avenue
Pittsburgh, PA 15222
(412) 288-4713

General Research Laboratories
8900 Winnetka Avenue
Northridge, CA 91324
800-421-1856
Fax: (818) 407-8500

Gero Vita
6021 Yonge Street
Toronto, Ontario
Canada M2M 3W2
800-694-8366

Gero Vita International
520 Washington Street, #391
Marina Del Ray, CA 90292
800-825-8482

Global Nutritional Research Corp.
4022-1 S. 20th Street
Phoenix, AZ 85040
(602) 243-5189
Fax: (602) 243-6551

Good'N Natural Vitamins
90 Orville Drive
Bohemia, NY 11716
(516) 244-2041
Fax: (516) 224-2013

Green Foods Corporation
320 Oxnard Avenue
Oxnard, CA 90505
800-777-4430 or (805) 983-7470
Fax: (805) 983-8840

Health Enhancers, Inc.
8139 Corunne Road
Flint, MI 48532
800-792-9199
Fax: (810) 659-4949

Health Products Corp.
1060 Nepperhan Avenue
Yonkers, NY 10703
(914) 423-2900
Fax: (914) 963-6001

HealthSmart Vitamins
1921 Miller Drive
Longmont, CO 80501
800-492-3003

Henkel Corporation
5325 South Ninth Avenue
LaGrange, IL 60525
(708) 579-6150

Herbal Products and Development
P. O. Box 1084
Aptos, CA 95001
(408) 688-8706
Fax: (408) 688-8711

Highland Laboratories
110 South Garfield
Mt. Angel, OR 97362
800-547-0273

Holistic Animal Care
7334 E. Broadway
Tucson, AZ 85710
800-497-5665
Fax: (520) 886-1727

Infinity2 Distribution, Inc.
63 E. Main Street, #700
Mesa, AZ 85201
(602) 668-1856

Interior Design Nutritionals
75 West Center Street
Provo, UT 84601
(801) 345-2000
Fax: (801) 345-1999

International Enzyme Foundation, Inc.
P. O. Box 249, Highway 160
Forsyth, MO 65653
800-433-8589
Fax: (417) 546-6433

Jarrow Formulas
1824 S. Robertson
Los Angeles, CA 90035
800-726-0886
Fax: (213) 204-2520

J. R. Carlson Laboratories
15 College
Arlington Heights, IL 60004
800-323-4141
Fax: (708) 255-1605

Juice Plus+
NSA
4260 E. Raines Road
Memphis, TN 38118-6977
(901) 366-9288

KAL, Inc.
P. O. Box 4023
Woodland Hills, CA 91365
(818) 340-3035

Klamath Blue Green, Inc.
P. O. Box 1626
Mt. Shasta, CA 96067
800-327-1956
Fax: (916) 926-6685

K. W. Pfannenschmit GmbH
P. B. 610151
22421 Hamburg
Germany
or
Habichthorst 36
22459 Hamburg
Germany
011-040-555-8660
Fax: 011-040-555-3898

Kyolic, Ltd.
23501 Madero
Mission Viejo, CA 92691
800-421-2998

Lactaid, Inc.
7050 Camp Hill Road
Fort Washington, PA 19034
800-LACTAID

Life Plus
P. O. Box 3749
Batesville, AR 72503
1-800-572-8446

Longevity Network, Ltd.
15 Cactus Drive
Henderson, NV 89014
(702) 454-7000
Fax: (702) 435-4786

Lotus Light Enterprises, Inc.
P. O. Box 1008
Silver Lake, WI 53170
(414) 889-8501

Magnus Enterprises
1406 West 178th Street
Gardena, CA 90248
(310) 532-8440
Fax: (310) 515-5263

MAK Wood Inc.
P. O. Box 184
Thiensville, WI 53092
(414) 242-2323
Fax: (414) 242-9448

Malabar Formulas
28537 Nuevo Valley Drive
Nuevo, CA 92567
800-426-6617

Marcor Development
108 John Street
Hackensack, NJ 07601
(201) 489-5700
Fax: (201) 489-7357

Marlyn Nutraceuticals
14851 North Scottsdale Road
Scottsdale, AZ 85254
800-462-7596

McNeil Consumer Products Co.
7050 Camp Hill Road
Ft. Washington, PA 19034
800-522-8243

Medi-Plex International
520 Washington Street, Suite 391
Marina del Rey, CA 90292
800-292-6006

Michael's Naturopathic Programs
6820 Alamo Downs Parkway
San Antonio, TX 78238
(210) 647-4700

Miller Pharmacal Group, Inc.
4562 Prime Parkway
McHenry, IL 60050
800-323-2915

Mucos Pharma GmbH & Co.
Malvenweg 2
D-82538 Geretsried
Germany
011-49-0-8171-5180
Fax: 011-49-0-8171-52008

Murdock Madaus Schwabe
10 Mountain Springs Parkway
Springville, UT 84663
(801) 489-1413
Fax: (810) 489-1700

National Enzyme Company
P. O. Box 128
Forsyth, MO 65653
800-433-8589
Fax: (417) 546-6433

National Vitamin Company, Inc.
2075 West Scranton Avenue
Porterville, CA 93257
(209) 781-8871
Fax: (209) 781-8878

Natren
3105 Willow Lane
Westlake Village, CA 91361
800-992-3323

Natrol, Inc.
20731 Marilla Street
Chatsworth, CA 91311
800-326-1520
Fax: (818) 701-0623

Naturally Vitamin Supplements Co.
14851 N. Scottsdale Road
Scottsdale, AZ 85254
800-899-4499
Fax: (602) 991-0551

Nature's Concept
5242 Bolsa Avenue, Suite 3
Huntington Beach, CA 92649
(714) 893-0017
Fax: (714) 897-5677

Nature's Life
7180 Lampson Avenue
Garden Grove, CA 92841
(714) 379-6500
Fax: (714) 379-6501

Nature's Plus
548 Broad Hollow Road
Melville, NY 11747
(516) 293-0030
Fax: (516) 249-2022

Nature's Products, Inc.
2525 Davie Road
Davie, FL 33317
800-752-7873
Fax: (305) 474-0989

Nature's Sunshine Products, Inc.
P. O. Box 19005
Provo, UT 84605-9005
(801) 342-4300

Nature's Way
10 Mountain Springs Parkway
Springville, UT 84663
800-926-8883
Fax: (801) 489-1700

Nebraska Cultures, Inc.
6610 Van Dorn
Lincoln, NE 68506
(602) 230-2758

Nevada Nutritional
4900 Mill Street, #B-32
Reno, NV 89502
(702) 857-2700
Fax: (702) 857-1412

NF Formulas, Inc.
805 SE Sherman
Portland, OR 97214-4666
800-547-4891

Nikken U.S.A., Inc.
15363 Barranca Parkway
Irvine, CA 92718
(714) 789-2000
Fax: (714) 789-2080

Novo Nordisk A/S
Nova Alle
2880 Bagsvaerd
Denmark
011-45-4444-8888
Fax: 011-45-4449-0555

Now Foods
550 Mitchell Road
Glendale Heights, IL 60139
800-999-8069 or (630) 545-9098

NuBotanic International, Inc.
15512 S. Figueroa Street
Gardenia, CA 90248
(310) 327-4500

NutriCology, Inc.
400 Preda Street
San Leandro, CA 94577
800-545-9960 or (510) 639-4572
Fax: (510) 635-6730

Nutri-Essence
100 N.W. Business Park Lane
Riverside, MO 64150
800-647-6377
Fax: (800) 844-1957

Nutrilabs, Inc.
5000 W. Oakey, #D12
Las Vegas, NV 89102
(702) 878-7376
Fax: (702) 878-4863

Nutritional Enzyme Support System (NESS)
2903 N.W. Platte Road
Riverside, MO 64150
800-637-7893

Nutri-West
P. O. Box 950
Douglas, WY 82633
(307) 358-5066

Oekpharma GmbH
Moosham 29
A-5580 Unterberg
Austria
011-43-6476-805-0
Fax: 011-43-6476-805-40

Optimal Nutrients
1163 Chess Drive, Suite F
Foster City, CA 94404
800-966-8874
Fax: (415) 349-1686

Pharmavite
15451 San Fernando Boulevard
Mission Hills, CA 91345
(818) 837-3633
Fax: (818) 837-6182

Phoenix Laboratories
140 Lauman Lane
Hicksville, NY 11801
(516) 822-1230
Fax: (516) 939-0234

PhysioLogics
6565 Odell Place
Boulder, CO 80301-3330
800-765-6775

PhytoPharmica
825 Challenger Drive
Green Bay, WI 54311
800-553-2370
Fax: (414) 469-4418

Phyto-Therapy, Inc.
P. O. Box 555
Franklin Lakes, NJ 07417
(201) 891-1104
Fax: (201) 848-1867

Prevail Corporation
2204-8 N.W. Birdsdale
Gresham, OR 97030
800-248-0885
Fax: (503) 667-4790

Professional Health Products
P. O. Box 80085
Portland, OR 97280-1085
800-952-2219
Fax: (503) 452-1239

Progressive Laboratories, Inc.
1701 W. Walnut Hill Lane
Irving, TX 75038
(214) 518-9660
Fax: (214) 518-9665

Progressive Research Labs, Inc.
9396 Richmond, Suite 514
Houston, TX 77063
800-877-0966

Quad Laboratories
P. O. Box 555
Franklin Lakes, NJ 07417
(201) 891-1104
Fax: (201) 848-1867

Quest Vitamins
5180 S. Service Road
Burlington, Ontario
Canada L7L 5H4
(905) 637-7800

Rainbow Light Nutritional Systems
P. O. Box 600
Santa Cruz, CA 95061
800-635-1233

Really Raw Honey
1301 S. Baylis Street
Baltimore, MD 21224
(410) 675-7233
Fax: (410) 675-7411

Rexall Showcase International
851 Broken Sound Parkway, NW
Boca Raton, FL 33487
(561) 994-2090
Fax: (561) 241-5319

Rexall Sundown, Inc.
851 Broken Sound Parkway, NW
Boca Raton, FL 33487
(407) 241-9400
Fax: (407) 995-6880

The Rockland Corporation (TRC)
12320 E. Skelly Drive
Tulsa, OK 74128
(918) 437-7310 or (918) 437-7311

Sabinsa Corporation
121 Ethel Road West, Unit #6
Piscataway, NJ 08854
(732) 777-1111
Fax: (732) 777-1443

Savin Your Health Products
A division of Enzyme Process NY, Inc.
P. O. Box 30027
Staten Island, NY 10303
800-762-6841 or (718) 494-8446
Fax: (718) 370-2942

Schiff Products, Inc.
1960 South 4250 West
Salt Lake City, UT 84104
(801) 975-5000
Fax: (801) 972-6532

Set-N-Me-Free Aloe Vera Co.
19220 S.E. Stark
Portland, OR 97233
800-221-9727
Fax: (503) 669-9057

Sigma Chemical Co.
P. O. Box 14508
St. Louis, MO 63178
800-325-3010
Fax: (314) 771-5757

Solaray, Inc.
1104 Country Hill Drive, Suite 412
Ogden, UT 84403
(801) 626-4956
Fax: (801) 393-8215

Solgar Vitamin and Herb Co.
500 Willow Tree Road
Leonia, NJ 07605
800-645-2246
Fax: (201) 944-7351

Source Naturals
P. O. Box 2118
Santa Cruz, CA 95063
(408) 438-1144
Fax: (408) 438-7410

Specialty Enzymes and Biochemicals Co.
5390 La Crescenta
Yorba Linda, CA 92687
(714) 692-3350
Fax: (714) 692-3051

Standard Process
12521 131st Court, NE
Kirkland, WA 98083-2484
800-292-6699

Thompson Nutritional Products
4031 N.E. 12th Terrace
Ft. Lauderdale, FL 33334
800-421-1192

Tishcon Corporation
30 New York Avenue
Westbury, NY 11950
800-848-8442
Fax: (516) 997-3660

Triarco
6 Morris Street
Paterson, NJ 07501
(201) 278-7300
Fax: (201) 278-0317

Twin Laboratories, Inc.
2120 Smithtown Avenue
Ronkonkoma, NY 11779
(516) 467-3140
Fax: (516) 471-2375

Tyler Encapsulations
2204-9 NW Birdsdale
Gresham, OR 97030
800-869-9705

Valley Research, Inc.
P. O. Box 750
South Bend, IN 46624-0750
(219) 232-5000
Fax: (219) 232-2468

Viobin Corporation
700 E. Main Street
P. O. Box 158
Waunakee, WI 53597
(608) 849-5944

Vitagenics
240 South Broad Street
P. O. Box 886
Elkhorn, WI 53121
(414) 723-4942
Fax: (414) 723-5462

Vitamin Research Products, Inc.
3579 Highway 50 East
Carson City, NV 89701
800-877-2447
Fax: (702) 844-1331

Vita-Pure, Inc.
410 W. 1st Avenue
Roselle, NJ 07203
(908) 245-1212
Fax: (908) 245-1999

Wakunaga of America Co., Ltd.
23501 Madero
Mission Viejo, CA 92691
(714) 855-2776
Fax: (714) 458-2764

Weider Food Companies
1911 South 3850 West
Salt Lake City, UT 84104
(801) 972-0330

Westar Nutrition, Inc.
1239 Victoria Street
Costa Mesa, CA 92627
(714) 645-6100
Fax: (714) 645-9131

Wild Rose Herbal Formulas
#203, 8173-128th Street
Surrey, British Columbia
Canada V3W 4G1
(604) 591-8881
Fax: (604) 597-1784

Y. H. Products Corporation
400 North Lombard Street
Oxnard, CA 93030
(805) 983-1130
Fax: (805) 983-3648

Zia Cosmetics
410 Townsend Street, 2nd Floor
San Francisco, CA 94107
(415) 543-7546

Bibliography

Adamson, I., et al., "Pepsins in Protein-Energy Malnutrition," *Enzyme* 39:44–49 (1988).

Alpers, David H. and Tedesco, Francis J., "The Possible Role of Pancreatic Proteases in the Turnover of Intestinal Brush Border Proteins," *Biochimica et Biophysica Acta* 401:28–40 (1975).

Alzheimer's Association. Retrieved from http://www.alz.org/facts/rtstats.htm.

ASPRS National Clearinghouse of Plastic Surgery Statistics press release, April 28, 1999.

Blonstein, J. L., "The Use of 'Buccal Varidase' in Boxing Injuries," *Practitioner* 185:78–79 (1960).

Borel, P., Armand, M., Senft, M., et al., "Gastric Lipase: Evidence of an Adaptive Response to Dietary Fat in the Rabbit," *Gastroenterology* 100:1582–1589 (1991).

Borgstrom, B., et al., "Studies of Intestinal Digestion and Absorption in the Human," *Jnl. Clin. Invest.* 36:1521–1536 (1957).

Burtis, G., Davis, J., Martin, S., *Applied Nutrition and Diet Therapy* (Philadelphia: W. B. Saunders Co., 1988).

Cancer Facts and Figures 1999: Basic Cancer Facts. Atlanta: American Cancer Society, 1999.

CDC National Center for Chronic Disease Prevention and Health Promotion, Retrieved from http://www.cdc.gov.nccdphp/arthritis.htm, 26 June 1999.

Centers for Disease Control. Retrieved from http://www.cdc.gov/niosh/diseas.html.

Centers for Disease Control, Chronic Disease Prevention. *Preventing Cardiovascular Disease: Addressing the Nation's Leading Killer At-A-Glance 1999*. Retrieved from http://www.cdc.gov/nccdphp/cvd/cvdaag.htm.

Centers for Disease Control, Office of Women's Health, *Sexually Transmitted Diseases*. Retrieved from http://www.cdc.gov/od/owh/whstd.htm.

Cichoke, A. J., *Acute Trauma and Systemic Enzyme Therapy* (Portland, OR: Seven C's Publishing, 1993).

Cichoke, A. J., *AIDS and Metabolic Therapy* (Portland, OR: Seven C's Publishing, 1992).

Cichoke, A. J., *A New Look at Enzyme Therapy* (Portland, OR: Seven C's Publishing, 1993).

Cichoke, A. J., *A New Look at Chronic Disorders and Enzyme Therapy* (Portland, OR: Seven C's Publishing, 1993).

Cichoke, A. J., *Neurologic Considerations in Toxic, Metabolic and Nutritional Disorders* (Portland, OR: Seven C's Publishing, 1990).

Cichoke, A. J., *New Hope for AIDS* (Portland, OR: Seven C's Publishing, 1993).

Cichoke, A. J., *Nutrition to Give Your Athlete the Winning Edge* (Portland, OR: Seven C's Publishing, 1990).

Cirelli, M. G., "Treatment of Inflammation and Edema with Bromelain," *Delaware Med. Jnl.* 34(6):159–167 (June, 1962).

Dewar, M. J. S., "New Ideas about Enzyme Reactions," *Enzymes* 36:8–20 (1986).

Eisenberg, D. M., et al., "Unconventional Medicine in the United States," *New Engl. Jnl. of Med.* 328(4):246–252 (1993).

Ellis, G., et al., editors, "Eighth International Congress of Clinical Enzymology, Toronto, Canada. May 30–June 1, 1989," *Clin. Biochem.* 22:401–415 (1989).

Endometriosis Association International. Retrieved from http://www.endometriosisassn.org/endoassn_main.htm.

FDA Consumer, Washington, D.C.: U.S. Food and Drug Administration. January/February 1998. Retrieved from http://vm.cfsan.fda.gov/~dms/.

Fields, M., et al., "Copper Uptake in vitro: Effect of Fructose or Glucose in the Medium," *Nutr. Rep. Intl.* 37(5):1117–1125 (1988).

Food and Drug Association. Retrieved from http://www.fda.gov/fdac/features/ 1998/598_pros.html.

Gardner, M. L. G., "Intestinal Assimilation of Intact Peptides and Proteins from the Diet—a Neglected Field?" *Biol. Rev.* 59:289–331 (1984).

Gardner, M. L. G., "Gastrointestinal Absorption of Intact Proteins," *Ann. Review of Nutr.* 8:329–350 (1988).

Gilboa, N., et al., "Inhibition of Fibrinolytic Enzymes by Thrombin Inhibitors," *Enzymes* 40:144–148 (1988).

Girard, L. J., "Dislocation of Cataractous Lens by Enzymatic Zonulolysis: A Suggested Solution to the Problem of the 18 Million Individuals Blind from Cataracts in Third-World Countries," *Ophthalmic Surgery* 26(4):343–345 (July–August 1995).

Giulian, B. B., et al., "Treatment of Pancreatic Exocrine Insufficiency. *In vitro* Lipolytic Activities of Pancreatic Lipase and Fifteen Commercial Pancreatic Supplements," *Ann. of Surg.* 165(4):564–570 (April, 1967).

Goldberg, D. M. and Okuda, K., editors, "Selected Papers from the 7th International Congress of Clinical Enzymology, Osaka, Japan. September 11–14, 1988," *Clin. Biochem.* 23(2):97–171 (1990).

Gonzalez, N. J., and L. L. Isaacs, "Evaluation of Pancreatic Proteolytic Enzyme Treatment of Adenocarcinoma of the Pancreas, with Nutrition and Detoxification Support," *Nutrition and Cancer* 33(2):117–124 (1999).

Graham, D. Y., "Pancreatic Enzyme Replacement," *Digestive Diseases and Sciences* 27(6):485–490 (June, 1982).

Grossman, M. I., Greengard, H., and Ivy, A. C., "The Effect of Dietary Composition on Pancreatic Enzymes," *Amer. Jnl. of Physiol.* 138:676–82 (1943).

Grossman, M. I., Greengard, H., and Ivy, A. C., "On the Mechanism of the Adaptation of Pancreatic Enzymes to Dietary Composition," *Amer. Jnl. of Physiol.* 41:38–41 (1944).

Guskov, A. R., I. D. Bogacheva, and G. B. Iatsevich, "Systemic (Vobenzyme Preparation) and Local Enzyme Therapy in Transurethral Drainage of the Prostate in Patients with Obstructive Forms of Chronic Prostatitis," *Urol. Nefrol.* 6:37–42 (November/December 1998).

Gyr, K., et al., "Effect of Oral Pancreatic Enzymes on the Course of Cholera in Protein-Deficient Vervet Monkeys," *Gastroenterology* 74:511–513 (1978).

Gyr, K., et al., "The Effect of Oral Pancreatic Enzymes on the Intestinal Flora of Protein-Deficient Vervet Monkeys Challenged with *Vibrio cholerae*," *Amer. Jnl. Clin. Nutr.* 32:1592–1596 (1979).

Hall, D. A., A. R. Zajac, R. Cox, and J. Spanswick, "The Effect of Enzyme Therapy on Plasma Lipid Levels in the Elderly," *Artherosclerosis* 43S: 209–215 (1982).

Health Care Financing Review, Washington, D.C.: U.S. Health Care Financing Administration, fall 1997. Retrieved from http://www.hcfa.gov/stats/stats.htm.

Heinicke, R. M., et al., "Effect of Bromelain (Ananase) on Human Platelet Aggregation," *Experientia* 28(7):844–845 (1972).

Hemmings, W. A. and Williams, E. W., "Transport of Large Breakdown Products of Dietary Protein Through the Gut Wall," *Gut* 19:715–723 (1978).

Hoefer-Janker, H., "The Importance of Vitamin A and Proteolytic Enzymes in Cancer Therapy," *Arztliche Praxis* 54:2805–2806 (1971).

Izaka, K., et al., "Gastrointestinal Absorpotion and Anti-Inflammatory Effect of Bromelain," *Japan Jnl. Pharmacol.* 22:519–534 (1972).

Jackson, P. G., et al., "Intestinal Permeability in Patients with Eczema and Food Allergy," *Lancet* i:1285–1286 (1981).

Kabacoff, B. L., et al., "Absorption of Chymotrypsin from the Intestinal Tract," *Nature* 199:815 (1963).

Keljo, D. J. and Hamilton, J. R., "Quantitative Determination of Macromolecular Transport Rate across Intestinal Peyer's Patches," *Amer. Jnl. Physiol.* 244:G637–644 (1983).

Khaw, K. T., et al., "Efficacy of Pancreatin Preparations on Fat and Nitrogen Absorptions in Cystic Fibrosis Patients," *Pediatr. Res.* 12:437 (1978).

Kierle, A. M., et al., "Chymotrypsin Dissolution of Intra-Arterial Thrombi by Perfusion of Isolated Extremity," *Arch. of Surg.* 81:159–174 (1960).

Kiessling, H. and Svensson, R., "Influence of an Enzyme from *Aspergillus oryzae*, Protease I, on Some Components of the Fibrinolytic System." *Acta. Chem. Scand.* 24:569–579 (1970).

Kligerman, A. E., "Relative Efficiency of a Commercial Lactase Tablet," *Amer. Jnl. Clin. Nutr.* 51:890–893 (1990).

Komatsu, W., K. Yagasaki, Y. Miura, and R. Funabiki, "Stimulation of Tumor Necrosis Factor and Interleukin-1 Productivity by the Oral Administration of Cabbage Juice to Rats," *Bioscience, Biotechnology, and Biochemistry* 61(11):1937–1938 (November 1997).

Larsson, L. J., et al., "Properties of the Complex between α_2-Macroglobulin and Brinase, a Proteinase from *Aspergillus oryzae* with Thrombolytic Effect," *Thrombosis Research* 49:55–68 (1988).

Laskowski, M., et al., "Effect of Trypsin Inhibitor on Passage of Insulin across the Intestinal Barrier," *Science* 127:1115–1116 (1958).

Lee, K. K., Larsen, R. D., and Posch, J. L., "Evaluation of an Oral Proteolytic Enzyme in Operations Upon the Hand," *Surgery, Gynecology & Obstetrics* 595–597 (1967).

Levitt, Joseph A., Director, Center for Food Safety and Applied Nutrition, Food and Drug Administration, Department of Health and Human Services, testimony before the Committee on Government Reform, U.S. House of Representatives, May 27, 1999.

Liebow, C. and Rothgman, S. S., "Enteropancreatic Circulation of Digestive Enzymes," *Science* 189:472–474 (1975).

Liu, J. N., Y. Yoshida, M. Q. Wang, Y. Okai, and U. Yamashita, "B Cell Stimulating Activity of Seaweed Extracts," *International Journal of Immunopharmacology* 19(3):135–142 (March 1997).

Loehry, C. A., et al., "Permeability of the Small Intestine to Substances of Different Molecular Weight," *Gut* 11:466–470 (1970).

Lund, F., et al., "Thrombolytic Treatment with I. V Brinase of Advanced Arterial Obliterative Disease of the Limbs," *Angiology* 26:534–556 (1975).

Lynch, S. M. and Strain, J. J., "Effects of Dietary Copper Deficiency on Hepatic Antioxidant Enzymes in Male and Female Rats," *Nutr. Reports. Intl.* 37(5):1127–1131 (1988).

Malagelada, Juan R., Go, Vay L. W., and Summerskill, W. H. J., "Different Gastric, Pancreatic, and Biliary Responses to Solid-Liquid or Homogenized Meals," *Digestive Disease and Sciences* 24(2):101–110 (1979).

Martynenko, V., "Wobenzym in the Combined Pathogenetic Therapy of Chronic Urethrogenic Prostatitis," *Lik. Sprava* 6:118–120 (August 1998).

Masson, M., "Bromelain in Blunt Injuries of the Locomotor System. A Study of Observed Applications in General Practice," *Forschr Med* 113(19):303–306 (10 July 1995).

Matthews, D. M., "Memorial Lecture: Protein Absorption—Then and Now," *Gastroenterology* 73(6):1267–1279 (1977).

Michael, J. G., "The Role of Digestive Enzymes in Orally Induced Immune Tolerance," *Immunological Investigaitons*, 18 (9 & 10):1049–1054 (1989).

Midda, M., and M. W. Cooksey, "Clinical Uses of an Enzyme-Containing Dentifrice," *Journal of Clinical Periodontology* 13(10):950–956 (November 1986).

Miechowski, W. I. and Ercoli, N., "Studies on Proteolytic Enzymes, II, Trypsin and Chymotrypsin in Relation to Inflammatory Processes," *Jnl. Pharmacol and Exper. Ther.* 116:43–44 (1956).

Moore, F. T., "Some Further Views on Alpha-Chymotrypsin in Facial Surgery," *Brit. Jnl. of Plastic Surgery* 16:387–390 (1963).

Moss, D. W., "Clinical Enzymology—a Perspective," *Enzyme* 25:2–12 (1980).

Nassif, E. G., et al., "Brief Clinical and Laboratory Observations," *The Jnl. of Pediatrics* 98:320–323 (February, 1981).

National Multiple Sclerosis Society Information Resource Center and Library, Compendium of Multiple Sclerosis Information, 1997.

National Vital Statistics Report, Vol. 47, No. 9, Washington, D.C.: Centers for Disease Control, November 10, 1998.

Pache, Th. and Reichmann, H., "On the Stability of Key Enzymes of Energy Metabolism in Muscle Biopsies," *Enzyme* 43:183–187 (1990).

Polgar, Laszlo, *Mechanisms of Protease Action* (Boca Raton, Florida:CRC Press, Inc., 1989).

Potter, Van Renselaer, "Studies on Enzyme Inhibition—50 Years Ago," *FASEB Jnl.* 7(5):486–487 (1993).

Rammer, E., and F. Friedrich, "Enzyme Therapy in Treatment of Mastopathy. A Randomized Double-Blind Clinical Study," *Wiener Klinische Wochenschrift* 108(6):180–183 (1996).

Ransberger, Karl, "Die Enzymtherapie des Krebses," *Sonderdruck* 1:22–34 (1989).

Ransberger, Karl and Wolf, Max, "Effect of Proteolytic Enzymes from Animal and Plant Origin Upon Growth Influence of Normal and Tumor Tissues and Appearance of Metastatic Frequency of Experimental Tumors," Presented before the Xth Intl. Cancer Congress, Houston, TX. (1970).

Reddy, K. N. N., "Streptokinase-Biochemistry and Clinical Application," *Enzymes* 40:79–89 (1989).

Remtulla, M. A., et al., "Is Chymotrypsin Output a Better Diagnostic Index than the Measurement of Chymotrypsin in Random Stool?" *Enzyme* 39:190–198 (1988).

Rodeheaver, G. T., et al., "Mechanisms by Which Proteolytic Enzymes Prolong the Golden Period of Antibiotic Action," *Amer. Jnl. of Surg.* 136:379–382 (1978).

Rosalki, S. D., "Genetic Influences on Diagnostic Enzymes in Plasma," *Enzymes* 39:95–109 (1988).

Rose, R. C. and Bode, A. M., "Tissue Mediated Regeneration of Ascorbic Acid: Is the Process Enzymatic?" *Enzymes* 46:196–203 (1992).

Schneider, M. U., M. L. Knoll-Ruzicka, S. Domschke, G. Heptner, and W. Domschke, "Pancreatic Enzyme Replacement Therapy: Comparative Effects of Conventional and Enteric-Coated Micropheric Pancreatin and Acid-Stable Fungal Enzyme Preparations on Steatorrhoea in Chronic Pancreatitis," *Hepatology and Gastroenterology* 32S:97–102 (1985).

Schwimmer, Sigmund. *Source Book of Food Enzymology.* Westport, Conn.: The AVI Publishing Co., Inc., 1981, 652.

Seligman, B., "Bromelain: An Anti-Inflammatory Agent," *Angiology* 13:508–510 (1962).

Seltzer, A. P., "Minimizing Post-Operative Edema and Ecchymoses by the Use of an Oral Enzyme Preparation (Bromelain)," *E.E.N.T. Monthly* 41:813–817 (1962).

Shape Up America! Press release. Retrieved from http://www.shapeup.org/surveys/barrier.htm.

Shigei, T., Akira, S., and Tsune, N., "A Study on the Protective Effect of Bromelain, Crude Pineapple Proteases, Against Adrenaline Pulmonary Edema in Rats," *Japanese Heart Jnl.* 8(6):718–720 (1967).

Singer, F., and H. Oberleitner, "Drug Therapy of Activated Arthrosis. On the Effectiveness of an Enzyme Mixture versus Diclofenac," *Wiener Medizin Wochenschrift* 146(3):55–58 (1996).

Smolina, T. P., T. G. Orlova, O. N. Shcheglovitova, and N. N. Besednova, "Interferon-Inducing Action of Polysaccharide-Containing Biopolymers from Ginseng Root and Cell Culture," *Antibiotikii Khimioterapiia* 43(11):21–23 (1998).

Smyth, R. D., et al., "Systemic Biochemical Changes Following the Oral Administration of a Proteolytic Enzyme, Bromelain," *Arch. Int. Pharmacodyn.* 136:230–236 (1962).

Taussig, S., "The Mechanism of the Physiological Action of Bromelain," *Med. Hypotheses* 6:99–104 (1980).

Tylewska, S., S. Tyski, and W. Hrynie-Wicz, "The Effect of S. Aureus Lipase on Granulocyte Chemotaxis," *Medycyna Doswiadczalna I Mikrobiologia* 35S:171–174 (1983).

U.S. Food and Drug Administration. Retrieved from http://www.fda.gov/fdac/features/1998/598_pros.html.

Vail, D., "Report of the Committee on use of alpha-chymotrypsin in ophthalmology," *Trans. Am. Acad. Opth. and Otol.* 64:16–36 (1960).

Vamos, E. and Liebaers, I. "Prenatal Diagnosis of Inborn Errors of Metabolism," *Enzyme* 32:47–55 (1984).

VanHove, P., et al., "Action of Brinase on Human Fibrinogen and Plasminogen," *Thrombos. Haemostas.*, 42:571–581 (1979).

Verdery, R. B., D. K. Ingram, G. S. Roth, and M. A. Lane, "Caloric Restriction Increases HDL2 Levels in Rhesus Monkeys (*Macaca mulatta*)," *American Journal of Physiology* 273(4pt 1):E714–719 (October 1997).

Verdery, R. B., and R. L. Walford, "Changes in Plasma Lipids and Lipoproteins in Humans During a 2-Year Period of Dietary Restriction in Biosphere 2," *Archives of Internal Medicine* 158(8):900–906 (27 April 1998).

Verhaeghe, R., Verstraete, M., et al., "Clinical Trial of Brinase and Anticoagulants as a Method of Treatment for Advanced Limb Ischemia," *Eur. Jnl. Clin. Pharmacol.* 16:165–170 (1979).

Walker, W. A. and Isselbacher, K. J., "Uptake and Transport of Macromolecules by the Intestine," *Gastroenterology* 67(3):531–550 (1974).

Warshaw, A. L., et al., "Protein Uptake by the Intestine: Evidence for Absorption of Intact Macromolecules," *Gastroenterology* 66:987–992 (1974).

Wilson, R., "Plant Enzymes Can Make Your Skin Healthier," *Let's Live* (April):61–64 (1993).

Winchester, B., "Prenatal Diagnosis of Enzyme Defects," *Arch. of Disease in Childhood* 65:59–67 (1990).

Wolf, J. L. and Bye, W. A., "The Membranous Epithelial (M) Cells and the Mucosal Immune System," *Ann. Rev. Med.* 35:95–112 (1984).

INDEX

Page locators followed by the letter "t" indicate a table

A
Absorption
  problems with, 27, 30
    biochemical, 33–41
    mechanical, 32–33
    stress and, 41–42
  in small intestine, 25–26
  of vitamin A, 67
Acetylcholine, 48
Achlorhydria, 34–35
Acidity, 18–19. *See also* Hydrochloric acid
Acquired immune deficiency syndrome. *See* AIDS
Adenosine diphosphate. *See* ADP
Adenosine triphosphatase. *See* ATPase
Adenosine triphosphate. *See* ATP
Adnexitis, 198–199
ADP (adenosine diphosphate), 8
Aging
  cancer and, 120
  circulation problems and, 119, 212
  digestion and, 19, 44, 115–116
  energy needs and, 124–125
  enzymes and, 118–119
  exercise and, 128–129
  free radicals and, 116–118
  genetics and, 114–115
  immune system and, 116, 119–120
  lifespan, extending, 113–114, 129–130
  metabolic errors and, 115, 119–120
  of skin, 132–135
  weight control and, 121–128
AIDS (Acquired immune deficiency syndrome)
  definition of, 233
  enzyme therapy and, 162, 235–238
  immunosuppressive drugs and, 236
  long-term survival with, 234
  nutritional supplements and, 234–235
AIDS-related complex, 236
Alcohol and alcoholism
  B vitamins and, 73
  pancreatitis and, 36–38
  vitamin A and, 67–68
  vitamin C and, 71–72
  vitamin D3 and, 69
  zinc and, 77
Alkalinity, 18–19
Alkalizing punch, 60
Allergens, 153
Allergies, food, 134, 178
Alpha-galactosidase, 86
Aluminum, 116
Alzheimer's disease, 116
Amino acids, 14–15, 15*t*, 126–127
Amino peptidase, 29*t*
Amylases
  defined, 15
  gastric, 21, 28*t*
  pancreatic, 23, 29*t*, 125
  salivary (ptyalin), 16, 18, 28*t*
  supplemental, 92*t*, 94–95
Amylolytic enzymes. *See* Amylases
Anabolic enzymes, 263
Ananase. *See* Bromelain

Ankle injuries, 145–146
Ankylosing spondylitis (Bekhterev's disease), 167, 170–171
Antibodies
  antinuclear, 157
  autoantibodies, 157
  immune complexes and, 120, 161
Antigen mimicry, 120
Antigens, 120, 153
Antinuclear antibodies (ANA), 157
Antioxidants, 77–78, 117–118
Appetite control, 125–128
Arteriosclerosis, 208, 213, 214–215
Arthritis
  diet and, 167–168, 178
  enzyme therapy and, 162, 169, 175–177, 179
  five-step jump-start plus enzyme program for, 177–179
  "gouty," 174
  hypochlorhydria and achlorhydria in, 35
  immune complexes and, 159–160, 167, 169
  nonarticular rheumatoid, 171–172
  obesity and, 168
  osteoarthritis, 166, 172–173
  rheumatoid arthritis (*See* Rheumatoid arthritis)
Artificial kidney, 253
Aspergillus niger, 39, 43
Aspergillus oryzae, 38, 41, 43
Aspirin, 266
Astrup, T., 118
Atherosclerosis, 78, 208, 212
Athletic injuries. *See* Sports injuries
ATP (adenosine triphosphate), 7–9
ATPase (adenosine triphosphatase), 8
Attitude, positive mental, 109, 179
Autoantibodies, 157
Autoimmune diseases
  categories of, 157, 158*t*
  causes of, 160
  enzyme therapy for, 160–163
  immune complexes and, 157–161
  multiple sclerosis as, 241–242, 244
Azathioprine
  for arthritis, 175
  for multiple sclerosis, 243, 244

B
B cells, 154–155
B vitamins, 65, 72–73, 74*t*–75*t*, 75
Back
  injuries to, 146, 184–187
  pain in, 181–187
Bacterial prostatitis, 205
Baltimore, David, 225–226
Bartsch, W., 230
Basal metabolic rate (BMR), 9
Basal metabolism, 9
Bastyr College, 236
Baumuller, Marcel, 145–146
Beard, John, 162, 189, 257
Bekhterev's disease (ankylosing spondylitis), 167, 170–171
Benign prostatic hyperplasia, 205
Benitez, Helen, 189
Beta-carotene, 67–68
Bile, 26–27
Bile salts, 26–27
Bilirubin, 27
Bioflavonoids, 140
Biologic oxidation, 8
Biotin, 74*t*
Bland, Jeffrey, 91
Bliznakov, Emile, 75
Bloating, 86–87
Blood clots, 209–210, 212–214, 219
Blood coagulation
  aging and, 118–119
  in artificial kidney, 253
  in circulation, 209–212
  vitamin K and, 70
Blood-sugar (glucose), 3–4
BMR. *See* Basal metabolic rate
Boiling foods, 89
Boxing injuries, 144
Bradford Research Institute (BRI), 234
Breast diseases
  cancer, 194–195, 202–203
  fibrocystic or cystic, 200–202
Brighthope, Ian, 235
Bringmann, Wolfgang, 161
Bromelain
  activity of, 92*t*
  in injuries, 141
  in sinusitis, 249–250
  source and uses of, 86, 94
Bruckle, Dr., 175

C
Caffeine, 3, 106
Caillet, Rene, 181
Calories
  in food oxidation, 7, 8–9

life expectancy and, 122–123
obesity and, 183–184
requirements for, 9–10
Cancer
aging and, 120
breast, 194–195, 202–203
circulation disorders and, 220–221
diet and, 124
enzyme deficiencies and, 191
enzyme therapy and, 162–163, 189–190, 192–196
homeostasis and, 190–191
pancreatic, 193
prostate, 206–207
risks of developing, 188–189
skin, 188
tumor necrosis factor and, 192–193
viruses and, 226
vitamins and, 68, 71
Canning foods, 90
Capillaries, 137–140
Carbamate insecticides, 46
Carbohydrase, 39
Carbohydrates
amylolytic enzymes and, 15
complex, 11, 127–128
in diet, 4–5, 106
energy from, 8–9
sources of, 15
Carboxypolypeptidase, 24, 29*t*
Cardiovascular disease. *See* Atherosclerosis
Carotenoids, 67–68
Catabolic enzymes, 263
Catalysts, 5
Cataracts, 250–251
Celiac disease, 39
Cellulase, 15, 31–32, 92*t*, 95
Cellulose, 15, 17, 32
CF. *See* Cystic fibrosis
Chang, Thomas, M.S., 253
Chewing, 16–18, 31–33, 33*t*
Chicken pox virus, 227
Childbirth, 200
Children
lead poisoning in, 52
rheumatoid arthritis in, 169
Cholesterol
in circulation disorders, 208, 212–213
insulin and, 121
Cholesteryl ester hydrolase, 29*t*
Cholinesterase, 48–49
Chromium deficiency, 122
Chronic disorders. *See also specific disorders*
autoimmune diseases, 157–162
enzyme therapy in, 159–163
immune system in, 151–162
Chronic venous insufficiency, 216–217
Chyme, 21–25
Chymoral, 142
Chymotrypsin
in cataract surgery, 250
as a proteolytic enzyme, 15, 23, 24, 25*t*, 29*t*, 92*t*
supplemental, 93–94
Cichoke, Anthony, 142
Circulating immune complexes (CIC). *See also* Immune complexes
in ankylosing spondylitis, 170–171
in autoimmune disorders, 157
in multiple sclerosis, 244, 246
in rheumatic diseases, 169
Circulation
cholesterol in, 208, 212–213
disorders of
aging and, 119, 212
arteriosclerosis, 208, 214–215
cancer and, 220–221
chronic venous insufficiency, 216–217
lymphedema, 194, 202, 216, 220–221
pathologic venous processes, 217–219
post-thrombotic syndrome, 216, 219–220
thrombosis, 215–216
enzyme therapy in, 162, 213–214, 221–223
fibrin formation in, 210–212
homeostasis in, 208–210
Cirelli, M.G., 141
Citric acid cycle (Krebs cycle), 8
Cleeland, R., 226
Clenching teeth, 32
Clots, blood, 209–210, 212–214, 219
Coagulation
aging and, 118–119
in artificial kidney, 253
in circulation, 209–212
vitamin K and, 70
Coenzyme Q10, 75
Coenzymes
defined, 65–66
fat-soluble vitamins, 66–70
indirect, 65
Q10, 75
water-soluble vitamins, 70–75

Cofactors
defined, 65, 66
minerals, 76–83
Coley, William B., 192
Colitis, ulcerative, 159, 160
Collagen, 72, 133, 138
Colon
in digestive process, 26
function of, 57–58
Complement system, 158–159
Connective tissue, 183
Cooking foods, 88–90
Copper, 81–82
Cordaro, J.B., 105
Coronary heart disease. *See* Atherosclerosis
Corticosteroids, 148–149, 221, 243
Cortisone, 148, 243
Crohn's disease, 159, 160
Crosslinkages, 117–118
Culbert, Michael, 234
Cystic breast disease, 200–202
Cystic fibrosis (CF), 35–36
Cytostatic medications, 175
Cytotoxic T cells, 154

D
Davis, Adelle, 4
Death, causes of, 102, 106
Degenerative joint disease, 166, 172–173
Denck, H., 218
Dental plaque, 251–252
Deoxyribonuclease, 24, 29*t*
Desser, Lucia, 192–193
Detoxification
alkalizing punch, 60
in five-step jump-start enzyme program, 104
goals of, 55–56, 57–58, 60–61
kidney flush, 59
liver and gallbladder flush, 58–59
need for, 56–57
purge, 60
retention enema, 60
for rheumatic diseases, 177–178
Diclofenac, 173
Diet
aging and, 124–125
Alzheimer's disease and, 116
arthritis and, 167–168, 178
do's and don'ts in, 107
energy and, 3–5, 7–11
enzymes in, 42–43
in five-step program, 105–107, 178
fresh and raw foods in, 42, 85–88
in injuries, 141
life expectancy and, 122–123
multiple sclerosis and, 243, 245
processed foods in, 87–90, 106
skin and, 134
sugar and carbohydrates in, 4–5, 106
supplemental enzymes in, 91–96
for weight loss, 127–128
Digestion
absorption of foods in, 25–26, 27, 30, 32–42
aging and, 19, 44, 115–116
chewing in, 16–18, 31–33, 33*t*
enzyme therapy and, 42–44, 91
gastrointestinal tract in
jejunum and ileum in, 24–26
large intestine in, 26
liver in, 26–27
pancreas in, 22–24, 25*t*
purpose of, 13–14
saliva in, 16, 18, 28*t*, 32–33
small intestine in, 21–24
stomach in, 19–21
pH balance in, 18–19
problems with
biochemical, 33–41
mechanical, 31–33
stress, 32, 41–42
Digestive enzymes. *See* Enzymes, digestive
Dihydrofolate reductase, 73, 75
Dipeptidases, 29*t*
Disease prevention, xv–xvi, 102–103, 258
Diseases of civilization, 183–184
Dittmar, Friedrich-Wilhelm, 198–200
Dorrer, Dr., 229
Drugs
aspirin, 266
B vitamins and, 73, 75
corticosteroids, 148–149, 221, 243
immunosuppressive, 236
as inhibitors, 266
for multiple sclerosis, 243–244
penicillin, 253
for rheumatic diseases, 174–175
as toxins, 53
Dulbecco, Renato, 225–226
Duodenal ulcers, 23
Duodenum, 22
Dysmenorrhea, 199

E
Elastase, 23, 25*t*, 29*t*
Embolisms, 209
Endometriosis, 199
Enema, retention, 60
Energy
  ATP and, 7–9
  eating habits and, 4–5, 11
  enzymes and, xvi, 6–7
  free energy of food oxidation, 7
  metabolism of, 72–73
  needs for, 3, 124–125
  obtaining from nutrients, 7–10
  stress and, 4
  sugar and, 3–4
Enterically coated products, 35–36, 38
Enterokinase, 23–24
Enzyme cascades, 265
Enzyme deficiencies, 134, 191
Enzyme inhibitors, 87, 265–266
Enzyme therapy. *See also specific diseases or conditions*
  aging and, 118–119
  alternative to corticosteroids, 148–149, 221
  benefits of, xv–xvi
  cholesterol levels and, 213
  digestion and, 42–44, 91
  disadvantages of, 176–177
  dosage of, 147
  fibrinolytic, 214
  future of, 256–259
  immune complexes and, 159–161
  meals and, 147
  oral *vs.* injectibles, 146
  prophylactic use of
    in cancer, 193–194
    in injuries, 142–144, 146–147
    in venous disorders, 219
  side effects of, 176
  skin health and, 134–135
  with vitamins, 119, 195, 200–201, 203
Enzymes
  classification of, 261–262
  defined, 5
  diet and, 42–43
  effect of pesticides on, 48–50
  induced-fit theory of, 263
  inhibitors of, 87, 265–266
  molecular size of, 262
  naming, 261
  in nervous system, 10
  optimal pH of, 96*t*
  properties of, 264
  role in metabolism, 5–8
  sources of, 10–11
  specificity of, 6, 262–263
  substrates and, 262–265
  temperature's effect on, 88–90
Enzymes, digestive, 28*t*–30*t*
  categories of, 14–15
  gastric, 20–21
  pancreatic, 22–24, 25*t*, 125
  salivary, 16, 18, 28*t*
  supplemental, 43–44, 91–96
Enzymes, proteolytic. *See also* Enzyme therapy
  aging and, 119
  defined, 14–15
  digestive, 23–24, 25*t*
  effects on inflammation, 138–141, 139*t*
  in fruits and vegetables, 86
Enzymes, supplemental, 91–96. *See also* Enzyme therapy
  activities of, 92, 92*t*
  amylase, 92*t*, 94–95
  in arthritis, 179
  bromelain, 94
  cellulase, 95
  chymotrypsin, 93–94
  in digestion, 43–44
  lactase, 95
  lipase, 95
  pancreatin, 92–93
  papain, 94
  trypsin, 93
EPA (omega-3 fatty acid), 167
Ernst, A.M., 217
Excretion mechanisms, 57–58
Exercise
  aging and, 128–129
  arthritis and, 179
  in five-step jump-start enzyme program, 108
Extrusion cooking, 89–90

F
Fasting, 104–105
Fat-soluble vitamins, 66–70
  vitamin A, 67–69
  vitamin D_3, 69
  vitamin E, 69
Fatigue, 3–4
Fats
  energy from, 8–9
  lipases and, 15

Fatty feces (steatorrhea), 35, 38
Fiber, 15, 32, 128
Fibrin
  aging and, 119
  role in circulation, 210–212
Fibrinogen, 210
Fibrinolysis, 209–212
Fibrinolytic enzymes, 214
Fibrocystic breast disease, 200–202
Fischer, Professor, 262
Fitzgerald, Peter, 235
Five-Step Jump-Start Plus Enzyme Program
  detoxification, fasting, juicing, 103–105, 177–178
  exercise, 108, 179
  in improving lifestyle, 256–258
  for multiple sclerosis, 247
  nutrition and diet, 105–107, 178
  positive mental attitude, 109, 179
  for rheumatic diseases, 177–179
  supplements, 108, 179
Flatulence, 86–87
Folate, 73, 74*t*, 75
Food allergies
  in arthritis, 178
  enzymes and, 134
Food extracts, 87
Foods. *See also* Diet; Digestion
  absorption of, 25–26, 27, 30, 32–42
  in digestion, improving, 42–43
  enzymes in, 10–11, 127–128
  fresh or raw, 42, 85–88
  life expectancy and, 122–123
  obtaining energy from, 8–10
  processed, 87–90, 106
Football injuries, 142–143, 145
Free energy of food oxidation, 7
Free radicals
  aging and, 116–118
  arthritis and, 167
  cancer and, 190–191
  zinc and, 77–78
Frozen foods, 90
Fructose, 15
Fruits, 86, 88, 105, 107
Fulgrave, E.A., 141

G
Gall bladder flush, 58–59
Garlic, 87
Gas, 86–87
Gastric enzymes, 20–21, 28*t*
Gastric juices, 20–21, 27*t*
Gastric lipase, 20, 28*t*
Gastrointestinal (GI) tract
  detoxification of, 55–61
  jejunum and ileum in, 24–26
  large intestine in, 26
  liver in, 26–27
  pancreas in, 22–24, 25*t*
  purpose of, 13–14
  saliva in, 16, 18
  small intestine in, 21–26
  stomach in, 19–21
Gastrointestinal reflex, 23
Gebert, G., 43
Genetic coding
  aging and, 114–115
  multiple sclerosis and, 242
Genital herpes, 228–229
Gingivitis, 32, 252
Glenk, Wilhelm, 114, 194, 237
Glomerulonephritis, 159
Glucose, 3–4
Glucose Tolerance Factor (GTF) chromium, 122
Glutathione peroxidase, 82
Goebel, Professor, 170–171
Gold treatments, 175, 176
Gonzalez, N.J., 193
Gotz, H., 226
Gout, 173–174
Grains, 88
Green barley essence, 87
GTF chromium. *See* Glucose Tolerance Factor chromium
Gynecological diseases, 198–200

H
Hackman, Robert M., 80
Hagiwara, Yoshihide, 87
Heart disease, 102. *See also* Atherosclerosis
Heat value of food, 8–9
Heavy metals, 50–53
  detoxification of, 55
  lead, 51–52
  mercury, 52–53
Helper/inducer cells, 154
Hemorrhoids, 254
Herpes, 226–230
  genital (herpes simplex II), 228–229
  herpes zoster (shingles), 227–228, 229–230
High-jumping injuries, 145

HIV. *See* Human immunodeficiency virus
Holt, Henry T., 142
Homeostasis
  cancer and, 190–191
  in circulatory system, 208–210
  energy and, 11
Horger, I., 169
Hormone therapy, 201
Horrobin, David, 242
Howell, Edward, 42, 88
Human immunodeficiency virus (HIV)
  in AIDS definition, 233
  enzyme therapy and, 162, 235–238
  immunosuppressive drugs and, 236
  long-term survival with, 234
  nutritional supplements and, 234–235
Hutchinson-Gilford syndrome, 114–115
Hydrochloric acid
  hypochlorhydria and achlorhydria, 34–35
  production of, 19, 20
  secretin and, 23
Hydrolases, 261–262
Hydrolysis, 22
Hydrolytic enzymes
  in cancer treatment, 193
  in reducing immune complexes, 159
Hypochlorhydria, 34–35
Hypoglycemia, 4
Hypothalamus, 125–126

I
IC. *See* Immune complexes
Idiotypes, 120
Ileum, 24–26
Illness, prevention of, xv–xvi, 102–103, 258
Immune complexes (IC)
  aging and, 120
  in AIDS, 237
  arthritis and, 167, 169
  in autoimmune diseases, 157–161
  cancer and, 191
  circulating
    in ankylosing spondylitis, 170–171
    in autoimmune disorders, 157
    in multiple sclerosis, 244, 246
    in rheumatic diseases, 169
  enzyme therapy and, 159–161
  viruses and, 227
Immune system
  aging and, 116, 119–120
  anatomy of, 153–156
  autoimmune diseases, 157–162
  cells and secretions of, 156
  in chronic disorders, 152–162
  enzyme therapy and, 161–163
  self and nonself, 152–153
  stress and, 4
Immunosuppressive drugs, 236
Inderst, Rudolph, 220
Indirect coenzymes, 65
Induced-fit theory of enzymes, 263
Infections, 161
Inflammation
  acute *vs.* chronic, 151
  nontraumatically caused, 147
  phases of, 136–138
  proteolytic enzymes in, 137, 138–141
Inflammatory membrane, 137
Inhibitors, enzyme, 87, 265–266
Injuries
  corticosteroids in, 148–149
  diet in, 141
  enzyme therapy in
    back injuries, 146, 184–187
    inflammation, 138–141, 147, 149
    preventing injuries, 146–147
    sports injuries, 141–146, 161–162
Insecticides, 46
Insulin resistance, 121–122
Intestines
  enzymes of, 29*t*–30*t*
  juices of, 21, 22, 27*t*
  large, 26
  maintaining function of, 57–61
  small, 21–22
Intrinsic factor, 21, 28*t*
Iron, 81
Irritable bowel syndrome, 42
Isaacs, L.L., 193
Ischemia, 144
Isomaltase, 30*t*
Isomerases, 262

J
Jager, Hans, 236
Janker Radiation Clinic, 194–195
Jaw problems, 17, 32, 33*t*
Jejunum, 24–26
Joint disease, degenerative, 166, 172–173
Juicing, 105, 107
Juvenile rheumatoid arthritis, 169

K
Karate injuries, 143–144
Keim, H., 220
Kidneys
  artificial, 253
  flush of, 59
Klein, Gert, 174, 175–176
Klein, K., 220
Kleine, Michael W., 147
Knees, 144, 145
Konig, W., 200
Koop, C. Everett, 123
Kugler, Hans, 114
Kunze, Rudolf, 161

L
Lactase
  in digestion, 23, 30*t*
  enzymatic activity of, 92*t*
  lactose intolerance and, 40–41
  supplemental, 95
Lactose
  in digestion, 15
  intolerance of, 40–41, 40*t*
Large intestine, 26, 27*t*
Lead, 51–52
Legumes, 88
Lichtmann, A.L., 144
Lien, Eric J., 53
Lifespan, extending
  eating less and, 122–123
  factors in, 113–114
  rules for, 129–130
Lifestyle, improving, xv–xvi, 256–258
Ligases, 262
Lin, David J., 190
Lipase inhibitors, 128
Lipases
  gastric, 20, 28*t*
  pancreatic, 23, 29*t*
  supplemental, 92*t*, 95
  vitamin A absorption and, 67
Lipid peroxidase reaction, 167–168
Lipolytic enzymes, 15, 23. *See also* Lipases
Liver
  in digestion, 26–27
  flush of, 58–59
Lumbar disc injuries, 146, 185–186
Lyases, 262
Lymph, 155
Lymph nodes, 155
Lymphangiosarcoma, 202
Lymphedema, 194, 202, 216, 220–221
Lymphocytes, 153–156
Lymphoid organs, 153–154
Lymphoid tissue, 155

M
Macrophages, 155
Maehder, Dr., 218
Magnes, G.D., 252
Magnesium, 80–81
Mahr, Herbert, 219
Maimonides, Moses, 108
Malabsorption, 27, 30
  biochemical problems, 33–41
  mechanical problems, 32–33
  stress and, 41–42
Maltase, 23, 29*t*
Manganese, 82
Marty, Leo, 142
Mastopathy, 200–201
Mayer, Jean, 183
McDougall, John, 240
Medications. *See* Drugs
Melancholia, 250
Melanoma, 188
Men
  prostate cancer in, 206–207
  prostatitis in, 205–206
Mercury, 52–53
Metabolism
  ATP in, 7–9
  basal, 9
  B vitamins and, 72–73
  defined, 5
  errors in, 115, 119–120
  obesity and, 125–126, 128
  role of enzymes in, 5–8
Metalloenzymes, 66. *See also* Minerals
Microminerals, 76
Micronutrient supplementation program, 235
Microthrombosis, 214, 215
Milk intolerance, 40–41, 40*t*
Minerals, 76–83
  copper, 81–82
  food processing and, 88, 90
  iron, 81
  magnesium, 80–81
  manganese, 82
  molybdenum, 82–83
  pesticides and, 49
  selenium, 82, 242

supplements of, 61, 108, 179
zinc, 76–80
Molybdenum, 82–83
Morl, H., 220
Moro, V., 169
MS. *See* Multiple sclerosis
Mucus, 21, 28*t*
Multiple sclerosis (MS), 240–247
causes of, 241–242
described, 241
diet in, 243, 245
Five-Step Jump-Start Plus Enzyme Program in, 247
immune complexes and, 244, 246
incidence of, 242
treatment of
azathioprine, 243, 244
cortisone, 243
enzyme therapy, 162, 244–247
Muscles, back, 182, 184
Myelin, 241

N
National Colleges of Naturopathic Medicine, 236
Nervous system
effect of heavy metals on, 51–52
effect of pesticides on, 48–50
enzymes in, 10
Neu, Sven, 114, 194, 237
Neuhofer, Christa, 240, 243, 245–246
Neurological disorders
enzyme treatments for, 250
multiple sclerosis, 162, 240–247
Neurotransmitters, 126–128
Niacin (B_3), 74*t*
Nonarticular rheumatoid arthritis, 171–172
Nonbacterial prostatitis, 205
Nonsteroidal anti-inflammatory drugs (NSAIDs), 174
Nucleases, 24
Nucleosidases, 30*t*
Nutrition. *See* Diet; Foods
Nutritional supplements. *See* Supplements, nutritional
Nuts, 87

O
Obesity
arthritis and, 168
causes of, 127
disease and, 121, 129
insulin resistance and, 121–122
life expectancy and, 122–123
metabolism in, 125–126, 128
physical activity, calories and, 183–184
strategies for weight loss, 127–128
Organ-specific/nonspecific disorders, 157, 158*t*
Organophosphorus insecticides, 46
Osteoarthritis, 35, 166, 172–173
Osteoarthrosis, 173
Overweight. *See* Obesity
Oxalates, 78
Oxidoreductases, 261

P
Pageler, John, 243
Pain
arthritis and, 167
back, 181–187
chewing and, 32
after childbirth, 200
chronic venous insufficiency and, 216–217
herpes zoster and, 227–228
pancreatitis and, 37
sports injuries and, 141–146
thrombosis and, 215–216
Pancreas
cancer of, 193
enzymes of
digestion and, 22–24, 25*t*, 28*t*–29*t*
obesity and, 125–126
pancreatic cancer and, 193
pancreatic insufficiency and, 253–254
production and release of, 38
vitamin A absorption and, 67
functions of, 22–24
juices of, 23–24, 27*t*
Pancreatic amylase, 23, 29*t*, 125
Pancreatic insufficiency, 24, 35, 253–254
Pancreatin, 92–93, 92*t*
Pancreatitis, 36–38
acute, 24, 37
chronic, 37–38
Pancreozymin, 23
Pantothenic acid, 75*t*
Papain *(Carica papaya)*
celiac disease and, 39
in dental surgery, 252
enzymatic activity of, 92*t*
sources of, 86, 94
in sports injuries, 142
Papayas, 86, 86*t*

Pasteurization, 89
Pathologic venous processes, 217–219
Pelvic inflammatory disease (PID), 198
Pendular movements of intestine, 22
Penicillamine, 175
Penicillin, 253
Penicillinase, 253
Pepsin (gastric protease), 20, 28*t*, 34, 43
Pepsinogen, 20
Pesticides
 benefits and dangers of, 46–48
 controlling use of, 50
 effect on enzymes, 48–50
pH
 balance of, 18–19
 of common substances, 18*t*
 optimal enzyme, 96*t*
Phagocytes, 154
Phenylketonuria, 34
Phlebitis, 218
Phosphatase, 30*t*
Phospholipase A, 23, 29*t*
Phytase, 89–90
Phytates, 78, 89–90
Pineapples, 86
Pinocytosis, 25–26, 43, 91
Plaque, dental, 251–252
Plasmin
 activation of, 119, 214
 in arteriosclerosis, 214
 in coagulation, 118–119, 211–212
Plasminogen, 118–119, 211
Polynucleotidase, 30*t*
Positive mental attitude, 109, 179
Post-thrombotic syndrome, 216, 219–220
Priestley, Joan, 109, 234–235
Processed foods, 87–90, 106
Proctology, 254
Prophylactic enzyme therapy
 in cancer, 193–194
 in injuries, 142–144, 146–147
 in venous disorders, 219
Prostaglandins, 174, 242
Prostate cancer, 206–207
Prostate gland, 205–207
Prostate-specific antigen (PSA) test, 206
Prostatitis, 205–206
Prostatodynia, 205
Proteases. *See* Enzymes, proteolytic
Protein
 chains of, 117–118
 in diet, 106
 energy from, 8–9
 metabolism of, 27
 proteolytic enzymes and, 14–15
Proteolytic enzymes. *See* Enzymes, proteolytic
Prothrombin, 211
PSA test. *See* Prostate-specific antigen test
Ptyalin (salivary amylase), 16, 18, 28*t*
Pulmozyme (dornase alfa), 36
Purge, 60

R
Rahn, Hans-Dieter, 144
Ramirez, G., 235
Ransberger, Karl, 119
Rathgeber, W.F. (Rob), 142
Reaven, Gerald M., 121
Regulatory T cells, 154
Rennin, 21, 28*t*
Resnick, Corey, 134
Respiratory viruses, 230
Retention enema, 60
Retinol, 67–68
Retinol-binding protein, 67
Retinol dehydrogenase, 68
Rheumatic diseases
 ankylosing spondylitis, 167, 170–171
 categories of, 168
 drug therapy for, 174–175
 enzyme therapy in, 162, 169, 170–171, 175–177, 179
 five-step jump-start plus enzyme program for, 177–179
 gout, 173–174
 immune complexes and, 159–160
 osteoarthritis, 35, 166, 172–173
 osteoarthrosis, 173
 rheumatoid arthritis (*See* Rheumatoid arthritis)
 soft-tissue rheumatism, 171–172
Rheumatism. *See* Rheumatic diseases
Rheumatoid arthritis
 causes of, 167–168
 enzyme therapy and, 162, 169
 hypochlorhydria and achlorhydria, 35
 immune complexes and, 159–160
 juvenile, 169
 nonarticular, 171–172
Riboflavin (B_2), 74*t*
Ribonuclease, 24, 29*t*
Rokitansky, Ottokar, 193, 202, 203
Running injuries, 145

S
Saliva, 16, 32–33
Salivary enzymes, 16, 18, 28*t*
Scheef, Wolfgang, 200–201, 220
Secretin, 23
Sedivi, Dr., 244
Seeds, 87
Selenium, 82, 242
Self and nonself in immune system, 152–153
Self-tolerance, 153
Serotonin, 126–128
Shingles, 227–228, 229–230
Side effects
  of enzymes, 176
  of penicillin, 253
Sinusitis, 249–250
Skin
  aging of, 132–135
  cancer of, 188
  loss of elasticity in, 117, 134
SLE. *See* Systemic lupus erythematosus
Slipped discs, 146, 185
Small intestine
  in digestion, 21–22
  enzymes of, 29*t*–30*t*
  secretions of, 27*t*
SOD. *See* Superoxide dismutase
Soft-tissue rheumatism, 171–172
Soy sauce, 42–43
Specificity of enzymes, 6, 262–263
Spleen, 155
Sports injuries
  enzyme therapy in, 141–146, 161–162
  in football, 142–143, 145
  in high-jumping, 145
  in karate, 143–144
  preventing, 146–147
  in running, 145
  in swimming, 144–145
Starches, 15, 38
Steatorrhea, 35, 38
Steffen, C., 160–161, 169
Stewart-Treves Syndrome, 202
Stimulants, 3, 106
Stinchfield, Frank E., 181
Stomach
  digestion in, 19–21
  enzymes in secretions of, 20–21, 28*t*
Streichhan, Peter, 220
Stress
  back pain and, 182–183
  digestion and, 32, 41–42
  energy levels and, 4
  physical, 161–162
Substrates, 262–265
Sucrase, 23, 29*t*
Sucrose, 15
Sugar, 3–4, 106
Superoxide dismutase (SOD), 78, 82
Supplemental enzymes. *See* Enzymes, supplemental
Supplements, nutritional
  AIDS and, 234–235
  for arthritis, 179
  in detoxification, 61
  in five-step jump-start plus enzyme program, 108
  of zinc, 80
Surgery
  for breast cancer, 194–195, 201–202
  for cataracts, 250–251
  dental, 252
  for hemorrhoids, 254
Swank, Roy, 240, 241
Swimming injuries, 144–145
Systemic lupus erythematosus (SLE), 157, 160

T
T cells, 116, 154–155
Taub, S.J., 249
Teeth
  chewing problems, 17, 32, 33*t*
  clenching (bruxing), 32
  dental plaque and decay, 251–252
  extraction of, 252
Temin, Howard, 225–226
Temperature's effect on enzymes, 88–90
Tension syndrome, 182–183
Thiamine (B_1), 73, 74*t*
Thiamine pyrophosphate, 73
Thrombi (clots), 209–210, 212– 214, 219
Thrombin, 211
Thromboembolic vascular diseases, 209
Thrombosis, 215–216
TNF. *See* Tumor necrosis factor
Toothpaste, 251–252
Toxins. *See also* Detoxification
  buildup of, 56–57
  drugs, 53
  heavy metals, 50–53
  pesticides, 46–50
Trace elements, 76
Transferases, 262
Trehalase, 30*t*

Trypsin
obesity and, 125
as proteolytic enzyme, 15, 24, *28t–29t*, *92t*
in sports injuries, 144
supplemental, 93
Trypsin inhibitor, 24
Tryptophan, 126–127
Tsiminakis, Professor, 244
Tumor necrosis factor (TNF), 192–193
Tumors. *See* Cancer
Tyrosine, 126–127

U
Uffelmann, Klaus, 172
Ulcerative colitis, 159, 160
Uric acid, 173

V
Valls-Serra, J., 218
Van Shaik, W., 169
Varicose veins, 216–217
Vasculitis, 213
Vegetables
in diet, 86–87, 88, 124
juicing, 105, 107
Venous diseases
chronic venous insufficiency, 216–217
lymphedema, 194, 202, 216, 220–221
pathologic venous processes, 217–219
post-thrombotic syndrome, 216, 219–220
thrombosis, 215–216
Vinzenz, Kurt, 252
Viruses, 225–231
enzymes and, 226, 229–231
herpes, 226–229
genital (herpes simplex II), 228–229
herpes zoster (shingles), 227–228
HIV, 233–238
in multiple sclerosis, 241
respiratory, 230
weak defense system and, 226
Vitamin A, 67–69, 203
Vitamin B_6, *74t*
Vitamin B_{12}, *74t*
Vitamin C, 71–72
Vitamin D_3 (cholecalciferol or calciol), 69
Vitamin E, 69, 119, 195, 200–201
Vitamin K, 69–70
Vitamins
fat-soluble, 66–70
vitamin A, 67–69, 203
vitamin D3, 69
vitamin E, 69, 119, 195, 200–201
vitamin K, 69–70
pesticides and, 49
supplements of, 108
water-soluble, 70–75
B vitamins, 65, 72–73, *74t–75t*, 75
vitamin C, 71–72

W
Walford, Roy, 122–123
Warfarin, 70
Warts, 230
Water-soluble vitamins, 70–75
B vitamins, 72–73, *74t–75t*, 75
vitamin C, 71–72
Weight control
appetite control and, 125–128
energy needs and, 124–125
insulin resistance and, 121–122
life expectancy and, 122–123
Werner syndrome, 114–115
Winfrey, Oprah, 125
Wisdom teeth extractions, 252
Wolf, Max, 119, 189, 218, 229, 244
Women
breast cancer in, 194–195, 202–203
fibrocystic or cystic breast disease in, 200–202
gynecological diseases in, 198–200
Worschhauser, Dr., 142
Wrba, Heinrich, 193

X
Xing-chang Ou, 53

Z
Zinc
as antioxidant, 77–78
factors affecting levels of, 78–79
functions of, 76–77
sources of, 79, 80
Zuschlag, J.M., 143